ELECTROSPINNING - MATERIAL, TECHNIQUES AND BIOMEDICAL APPLICATIONS

Edited by **Sajjad Haider** and **Adnan Haider**

Electrospinning - Material, Techniques, and Biomedical Applications
http://dx.doi.org/10.5772/62860
Edited by Sajjad Haider and Adnan Haider

Contributors
Fatma Yalcinkaya, Baturalp Yalcinkaya, Oldrich Jirsak, Ahmed Elzatahry, Jeong Hyun Yeum, Yawen Li, Therese Bou-Akl, Nongnut Sasithorn, Lenka Martinová, Jana Horáková, Rattanaphol Mongkholrattanasit, Jose Cornejo-Bravo, Luis Jesús Villarreal-Gómez, Aracely Serrano-Medina, Nuray Kizildag, Nuray Ucar

Published by InTech
Janeza Trdine 9, 51000 Rijeka, Croatia

Publishing Process Manager Romina Rovan
Technical Editor SPi Global
Cover InTech Design team

First published December, 2016
Printed in Croatia

Additional hard copies can be obtained from orders@intechopen.com

Electrospinning - Material, Techniques, and Biomedical Applications, Edited by Sajjad Haider and Adnan Haider
p. cm.
Print ISBN 978-953-51-2821-2
Online ISBN 978-953-51-2822-9

For EU product safety concerns:
IN TECH d.o.o., Prolaz Marije Krucifikse Kozulić 3, 51000 Rijeka, Croatia,
info@intechopen.com or visit our website at intechopen.com.

Meet the editor

Dr. Sajjad Haider is an assistant professor at the Chemical Engineering Department, King Saud University, Riyadh, Saudi Arabia, since May 2009. He received his MSc degree in 1999; MPhil in 2004 from the Institute of Chemical Sciences, University of Peshawar, KPK, Pakistan; and PhD degree in 2009 from the Department of Polymer Science and Engineering, Kyungpook National University, Daegu, South Korea. His research work focuses on the development of carbon nanotubes and biopolymer composites, polymer hydrogel actuators, and preparation of the electrospun nanofibers for environmental and biomedical applications.

Dr. Adnan Haider is a postdoctoral research fellow at Nano, Medical and Polymer Materials Department, School of Chemical Engineering, Yeungnam University, South Korea, since March 2016. He received his MSc degree in 2010 from Kohat University of Science and Technology, KPK, Pakistan and MS leading to PhD degree in 2016 from the Department of Polymer Science and Engineering, Kyungpook National University, Daegu, South Korea. His research work focuses on the development of scaffolds for tissue regeneration, biopolymer composites, polymer hydrogel, drug delivery systems, and preparing the electrospun nanofibers and assessing their potential application in removal of hazardous materials from aqueous medium and biomedical application.

Contents

Preface

With the advancement in nanotechnology, researchers are discovering new techniques for the fabrication of materials with a variety of morphologies. These morphologies are usually aimed at specific application. Among the various techniques reported in the literature, electrospinning has gathered significant interest due to its ability to fabricate nanostructures with unique properties such as high surface area, excellent pore interconnectivity, and flexibility with reasonable strength. Electrospinning has been the most widely used technique in the late twentieth (1990) and early twenty-first (2000) centuries. During this period, a number of new materials have been electrospun, but not much work is reported on the electrospinning setup. Recently, not only significant improvements have been made in the instrument design and nanomaterials produced but also new materials have been added to the list of electrospun materials. Thus, electrospinning, from both academic and technical viewpoints, is presently the most multipurpose technique for the fabrication of uniform random and aligned fibers. The importance of electrospinning can be imagined from the fact that almost all of the polymers (both synthetic and natural origin) and many ceramic materials have been subjected to electrospinning. The other most interesting aspect of electrospinning is that fiber can be tuned according to its application. The prominent applications of electrospun fibers include filtration, textiles, catalysis, and reinforcement and biomedical (in tissue engineering, drug delivery, and antibacterial study). Despite the extensive applications, electrospun fibers are produced at a low production rate when conventional needle electrospinning with single- or multi-jets setup is used, which hampers their mass production. Needless electrospinning technique is the most appropriate technique, which could be used for the mass electrospun fiber production. This book not only reviews the current status of the knowledge of electrospinning with a critical view on the achievements and novel perspectives but also discusses the emerging ideas and features, both from the East and West, with a focus on needless electrospinning. As Confucius said, it is gratifying when friends from different origins work for one purpose. The book has also comprehensively reported different material systems, techniques, fiber structures, fiber properties, and biomedical applications of the electrospun fibers.

Dr. Sajjad Haider
Chemical Engineering Department
College of Engineering, King Saud University
Riyadh, Saudi Arabia

Section 1

Electrospinning and Material

Recent Trends in Electrospinning of Polymer Nanofibers and their Applications as Templates for Metal Oxide Nanofibers Preparation

Moustafa M. Zagho and Ahmed Elzatahry

Additional information is available at the end of the chapter

http://dx.doi.org/10.5772/65900

Abstract

Scientists have been paying a special attention for the synthesis of one-dimensional (1D) morphologies to attain new phenomena and novel physicochemical characteristics of materials. Furthermore, 1D nanostructures exhibit long axial ratio, which has a great influence on the physical and chemical properties of materials. It is worth mentioning that electrospinning is one of the most common and efficient techniques used for the preparation of 1D polymer composite nanofibers. Using electrospinning, nanofibers were fabricated by electrostatic stretching of polymer viscous solution by applying a high voltage. This chapter discusses the synthesis of metal oxide nanofibers such as tin oxide (SnO_2), zinc oxide (ZnO), titanium oxide (TiO_2), and nickel oxide (NiO) using electrospinning process of polymer solution containing metal precursors and followed by annealing procedures to eliminate the polymer galleries, which were chosen as a sacrificial template for the preparation of metal oxide nanofibers. SEM, XRD, and XPS are equipped to characterize the electrospun metal oxide nanofibers and the results settle the formation of homogeneously distributed metal oxide nanofibers.

Keywords: electrospinning, metal oxide, nanofibers, SEM, XRD, XPS

1. Introduction

In the last decades, electrospinning attracted a vast portion of attention from both research and commercial points of view. Scientists paid great interest for electrospinning because of its ability to yield fibers in the submicron scale that are very hard to attain by handling standard mechanical fiber-spinning procedures [1–5]. Electrospinning is a technology that is broadly applied to fabricate polymer fibers with diameters ranging from 2 nm to several micrometers and is also known to have unique properties such as high surface area, controlled porosity, and mechanical strength [6–9]. The pore structure and the diameters of the produced fabrics and nanofibers can be controlled using this technique [10, 11]. Electrospun fibers have been successfully used in varied applications including protective clothing, biomedical, pharmaceutical, security, environmental engineering, chemical sensors, and electrode materials [12–19].

Recently, electrospinning has been used in the research of natural and synthetic polymer nanofibers [20] such as cellulose [21], polyurethanes [22], collagen [23] and hyaluronic acid [24]. Electrospinning can be handled to obtain large quantities of fibers using two-layer electrospinning scheme. This technique consists of a lower layer, which contains a ferromagnetic suspension, and an upper layer, which contains the polymer solution, and the multiple nozzles are arranged in lines or circles. One such example is the bubble electrospinning, which is demonstrated and developed by many researchers [25, 26]. Dosunmu et al. reached high production rates of fibers by using porous hollow tube to obtain multiple jets and by increasing the tube length and the number of holes [27]. In addition, coelectrospinning of different polymers has been reported to control the morphology and tailor the voids volume [28–30].

1.1. Electrospinning process

Electrospinning setup consists of a high voltage power supply, spinneret (like pipette tip), and grounded collected plates (like metal screen) and is conducted at room temperature. Polymer solution is charged by applying high voltage and then accelerated toward the collector, which is of opposite polarity [31, 32]. The polymer must be dissolved first before using in electro-

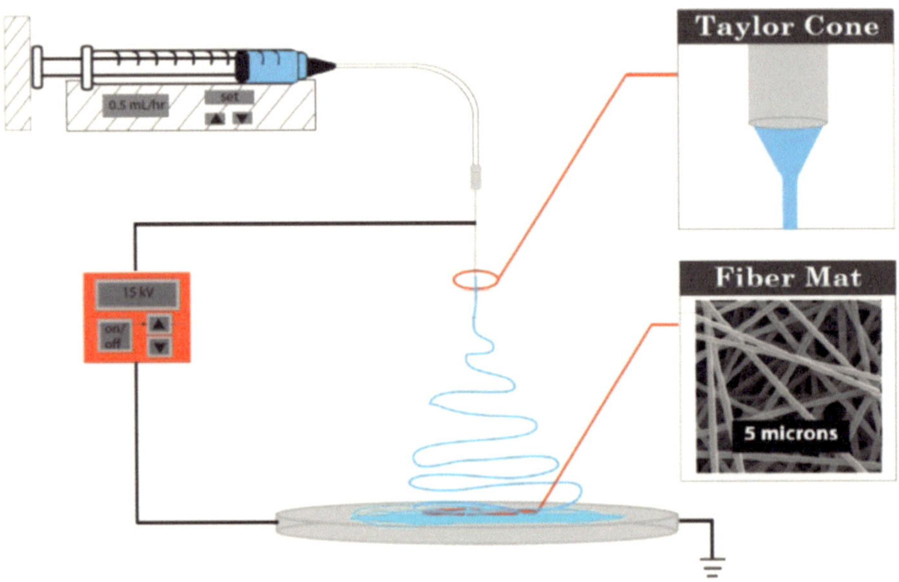

Figure 1. Schematic diagram of vertical electrospinning apparatus.

spinning and then introduced into the capillary tube. At critical voltage, the electrostatic forces counteract the surface tension of the used polymer solution and an electrified jet is produced and ejected from the tip of the Tylor cone and then the solvent will be evaporated and leaving a polymer nanofiber [33, 34]. Currently, there are two standard sets of electrospinning: vertical and horizontal. The vertical electrospinning is represented in **Figure 1**.

The diameters and morphology of electrospun nanofibers are measured by numerous parameters, which are classified into solution, process, and ambient parameters. These factors are demonstrated clearly in **Figure 2**.

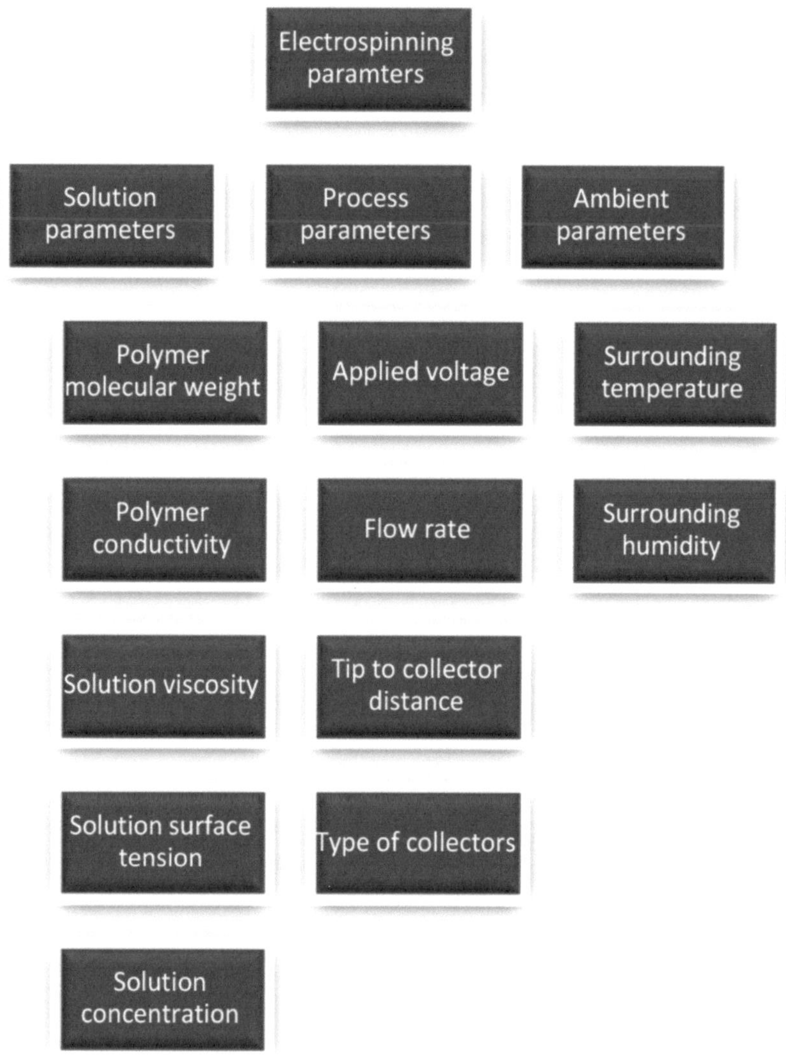

Figure 2. Factors affecting on elctrospinning technique.

2. Polymers favored in electrospinning technique

More than 200 polymers (natural, synthetic, and copolymers) were used in electrospinning successfully depending on their applications [35].

2.1. Natural polymers

Nowadays, polymer nanofibers are used in a wide variety of applications in various industries and also in other arenas. Natural polymers are preferred over synthetic polymers in medical and biological uses because of their low immunogenicity and better biocompatibility. Recent researchers have reported the use of electrospinning process of natural polymers such as gelatine, collagen, and silk fibroin [36–41]. It was mentioned that some natural polymers such as collagen lose their properties after electrospinning using fluoro alcohols [42]. To resolve this problem, highly volatile fluoro alcohol such as 2,2,2-trifluoroethanol is used during electrospinning of collagen [43]. Yang et al. [44] found that about 45% of collagen is lost during electrospinning process. In addition, beads formation represents another challenge for the preparation of natural polymer nanofibers. For instance, laminin I nanofibers are fabricated with beads in the mesh structure and this may be related to the presence of the cross-linking agent and these results are reported by Neal et al. [45].

2.2. Synthetic polymers

Synthetic polymers are favored over natural polymers in certain uses because synthetic polymers can be tailored to develop the mechanical and degradation properties [46]. Certain synthetic polymers can be used in biomedical applications with some limitations including polylactide (PLA) [47] and polyglycolide (PGA) [48] to produce electrospun nanofibrous scaffolds.

2.3. Copolymers

Copolymers appear as an attractive opportunity to obtain new structures with desirable characteristics from electrospinning and also have better efficiency than that of homopolymers. The mechanical, thermal, morphological, and biodegradability properties of polymers can be tailored by using copolymers in electrospinning process. Incorporation of the hydrophilic polymer of hydrophobic polyesters enhanced the cell affinity of this sort of polyesters. Moreover, the stiffness of poly (ethylene-co-vinyl alcohol) (PEVA) is improved after the addition of poly (glycolide) (PGA) [49]. Also, copolymerization of methyl methacrylate (MMA) with methacrylic acid (MAA) improved the thermal stability of poly methyl methacrylate [50]. In addition to that, electrospun scaffolds produced from poly (p-dioxanone-co-L-lactide)-block-poly (ethylene glycol) (PPDO/PLLA-B-PEG) have enhanced hydrophilicity and biodegradability due to the addition of PEG to PPDA-PLLA blends [51].

3. Metal oxide nanofibers

Metal oxide materials are critically important from both the academic and industrial point of views. Scientists have been paying a special attention for the synthesis of one-dimensional (1D) morphologies to attain new phenomena and novel physicochemical characteristics of materials.

3.1. Synthesis of crystalline metal oxide nanofibers and their applications

It is worth mentioning that electrospinning is the most common technique used for the preparation of 1D polymer composite nanofibers [52, 43]. Furthermore, 1D nanostructures exhibit long axial ratio, which has a great influence on the physical and chemical properties of materials. Using electrospinning, nanofibers are fabricated by electrostatic stretching of a viscous solution of polymer composites by applying a high voltage [3]. In the recent years, scientists have focused on the fabrication of metal oxide nanofibers such as tin oxide (SnO_2), zinc oxide (ZnO), titanium oxide (TiO_2), and nickel oxide (NiO) through the electrospinning of polymer solution containing metal precursors and followed by annealing procedures [53–57]. Depending on the thermal degradation of polymers, certain polymers such as polyvinyl alcohol (PVA) and polyacrylonitrile are chosen as a sacrificial template for the fabrication of metal oxide nanofibers. After the annealing process, the organic components will be degraded and the metals will be oxidized by inorganic precursors giving nanofibers the desired metal oxide.

3.1.1. Tin oxide (SnO_2) nanofibers

Recently, high interest has been paid to one-dimensional tin oxide (SnO_2) nanostructures. There are various techniques designed for the synthesis of different structures of SnO_2, such as electrospinning, self-catalytic, thermal evaporation, and thermal oxidation techniques [58–62]. The metal oxide fibers are commonly employed in gas detection application. The gas-sensitive characteristics of the prepared sensors are due to its high conductivity alterations when the prepared sensors adsorb or desorb a very low loading of chemical compounds on its surface [63–72]. High adsorption-desorption gas sensors in one-dimensional nanostructures were developed using metal oxide nanofibers owing to its highly porous networks [73, 74]. The morphological nature of the upper surface layer usually controls the molecular adsorption-desorption process, the response, and the sensitivity of the sensors. Metal oxide nanofibers are employed as a sensitive sheet instead of a thin film deposited by conventional processes such as drop, spinning, and airbrush because of its large surface area and, consequently, a thin film of metal oxide nanofibers enhances the sensitivity and the speed of the reaction.

Santos et al. [75] prepared SnO_2 nanofibers by electrospinning process of PVA/$SnCl_4.5H_2O$ composite followed by annealing at 450°C to eliminate PVA matrix. The fabricated SnO_2 electrospun nanofibers are designed as sensitive layers for resistive chemical sensors. The produced fibers exhibit high sensitivity, reproducibility, and quick reaction in the recognition of triacetone triperoxide (TATP) explosive precursors in the ppm scale. In addition, SnO_2 hollow nanofibers were fabricated for a high and fast detecting ethanol (C_2H_5OH) using polyvinylpyrrolidone (PVP) as a template, which was removed by calcination at 600°C for 2 h [76]. The produced SnO_2 nanofibers have a large surface area of 26.43 m^2g^{-1} and grain sizes of about 20 nm. Another study by Choi et al. [77] concerns the preparation of Pd-doped SnO_2 hollow nanofibers for methane (CH_4), H_2, carbon monoxide (CO), and C_2H_5OH detection. They studied the effect of temperature and Pd-doping content on the detection for CH_4, H_2, CO, and C_2H_5OH. They noticed a dramatic variance in the selectivity for these compounds between undoped and Pd-doped SnO_2 hollow nanofibers. The selective detection of C_2H_5OH was tested with undoped and 0.08 wt% Pd-doped SnO_2 hollow nanofibers. The study showed a significant decline in C_2H_5OH detection in case of using 0.4 wt% Pd when the temperature is raised from 330 to 440°C, while the responses to CH_4 and H_2 were improved and, consequently, they adjusted the selectivity of CH_4 and H_2 at 440°C with minimum interloping to C_2H_5OH. It was worth studying the enhancement of Pr-doping on the gas sensing performance due to its effective influence on the electronic and structural characteristics of materials. Li et al. [78] discussed the fabrication of Pr-doped SnO_2 hollow nanofibers for ethanol detection. They subjected a solution of PVP, $SnCl_2.2H_2O$, and $Pr(NO)_3.6H_2O$ for electrospinning. The prepared nanofibers sample was then annealed at 600°C for 2 h. It was reported that 0.6 wt% Pr-doped SnO_2 nanofibers gave the highest sensitivity and response for ethanol at 300°C. This may be related to the high surface area of the porous morphology and the increment in oxygen vacancy leading to enhanced oxygen absorption. Furthermore, Du et al. [79] focused on the fabrication of In_2O_3/SnO_2 nanofibers for formaldehyde gas sensors using bipolar electrospinning setup of double jets with opposite electric fields. Different sensors composition based on SnO_2, In_2O_3, and In_2O_3/SnO_2 nanofibers were tested at 375°C with 0.5–50 ppm of formaldehyde gas. The formaldehyde selectivity of In_2O_3/SnO_2 nanofibers is higher than that of SnO_2 and In_2O_3 nanofibers. The selectivity values for 0.5 and 50 ppm of formaldehyde were 2.2 and 18.9, respectively. In_2O_3/SnO_2 nanofibers have a good detection of formaldehyde in the interfere gasses of methanol, ethanol, acetone, and ammonia. In addition, Tang et al. [80] synthesized hollow hierarchical ZnO/SnO_2 nanofibers composite for methanol sensing applications. A solution of PVP, $SnCl_2.2H_2O$, and $Zn(NO_3)_2.6H_2O$ is used to prepare electrospun fibers, and then subjected to annealing at 600°C for 3h. This study noted that the nanofibers of ZnO/SnO_2 with a molar ratio of 1:1 exhibited an interesting methanol detection response at 350°C in the presence of interfering gasses.

3.1.2. *Zinc oxide (ZnO) nanofibers*

Another metal oxide of recent interest is ZnO, which is a chemically stable, nontoxic and inexpensive compound that exhibits high binding energy of 60 mV at 30°C and a direct band gap of 3.37 eV [81]. ZnO is industrially applied in a wide range of uses including chemical sensors, ultraviolet light-emitting diodes, dye-sensitized solar cells, functional instruments, piezoelectric materials, and transparent conductors [82–85]. The morphology, crystal structure, quality, and dimensions of ZnO control its electronic characteristics for different applications [86, 87].

The effective structures of 1D ZnO nanofibers have a small quantum restriction of charge carriers and this measure the performance of nanoscale devices [88–91]. Recent articles reported the synthesis of ZnO or ZnO-based nanofibers. Ding et al. [92] addressed the unique characteristics of ZnO nanofibers such as high hydrophobicity, development of nanograins, and electrical properties. The study reported the synthesis of super-hydrophobic ZnO nanofibers through a simple combination of wet chemical and electrospinning techniques. The produced ZnO nanofibers have been coated by fluoroalkyl silane (FAS) to obtain super-hydrophobic ZnO nanofibers. Because of the FAS modification, the fibrous ZnO films were transformed from super-hydrophilic (water contact angle (WCA) of 0°) to super-hydrophobic (WCA of 165°). Electrospinning technique was designed to produce nanofibers with a smooth surface which limits its use in some applications as the cohesion forces will be feeble, and thus rough surface nanofibers are preferred. In this context, Barakat et al. [93] mentioned that the electrospinning of colloidal solutions is useful for synthesizing rough surface ZnO nanofibers. The fibers were produced by electrospinning a colloidal solution of PVA, zinc nanoparticles (ZnNPs), and zinc acetate dehydrate (ZnAc) and then followed by calcination. Furthermore, Zhang et al. [94] addressed the fabrication of a hollow ZnO nanofibers using a ZnNPs-free solution. The study reported the preparation of ZnO hollow nanofibers with diameters of 120–150 nm using an electrospinnable solution of PAN, PVP, and ZnAc composites, then followed by calcination to remove the used polymers. They obtained core-shell structure from elctro-spinning process where PAN was the core and PVP/ZnAc composite was the shell. ZnO hollow nanofibers showed interesting results related to sensing characteristics against C_2H_5OH. Another study by Kanjwal et al. [95] reported the preparation of rough ZnO hierarchical nanofibers as a photocatalyst against methylene blue dehydrate. They electrospun a colloidal solution consisting of ZnNPs, PVA, and ZnAc dehydrate and then followed by a calcination step at 500°C. Moreover, electrospinning a mixture of high molecular weight PVP and ZnAc in dimethylformamide (DMF) synthesized 1D polycrystalline ZnO nanofibers with porous morphologies [96]. The study reported the effect of ZnAc loadings from 10 to 15 wt% after calcination process from 350 to 650°C onto the produced nanofibers. The prepared ZnO nanofibers were utilized for water treatment as a result of its photocatalytic performance.

A comparative study to compare the undoped and Ce-doped ZnO hollow nanofibers for acetone detection was carried by Li et al. [97]. The used electrospinning solution consisted of cerium nitrate, ZnAc, and PVP. The sample was then followed by calcination at 600°C for 1.5 h. The study declared that Ce-doped ZnO nanofibers are very sensitive to acetone at 260°C. This is ascribed to the morphological and electronic modification in the presence of Ce-doping. Moreover, Al-Ga co-doped ZnO nanofibers have attracted high interest because of its excellent cost effectiveness, conductivity improvement, and mechanical flexibility enhancement. Park and Han [98] reported the preparation and the characterization of Al-Ga co-doped ZnO nanofibers by electrospinning a solution of zinc acetate dihydrate, aluminum nitrate nonahydrate, gallium nitrate hydrate, and PVA and then by annealing at 550°C. The study showed a variation of the lattice parameter with the addition of Ga beyond 2 at.% Al-doping owing to smaller gaps in the atomic size. Consequently, more Ga content can be added without extensive strain. Al-Ga co-doped ZnO nanofibers with 2 at.% Al and 1 at.% Ga exhibit the highest conductivity of 9.57×10^{-3} Scm^{-1}. In addition, Ga incorporation to 2 at.% Al-ZnO nanofibers developed the mobility and the degree of crystallinity.

3.1.3. Titanium oxide (TiO_2) nanofibers

Wu et al. [99] addressed the usage of TiO_2 in many applications including solar, optoelectronic, and catalytic devices. TiO_2 exists in different morphologies such as rutile, anatases, brookite, TiO_2 B (bronze), and TiO_2 R (ramsdellite). It is worth mentioning that the most chosen TiO_2 crystal phase for lithium ion batteries is the anatase structure due to its high capacity and low-cost production [100–102].

He et al. [103] reported the fabrication of 100–300 nm conductive TiO_2 nanofibers using KOH treatment. In this study, a mixture solution of PVP and titanium tetraisopropanolate was used to produce nanofibers for supercapacitor electrode applications using electrospinning. The as prepared nanofibers exhibit high conductivity, high cycling stability, and high surface area with a stable specific capacitance even after 10,000 cycles and it is enhanced from 0.02 Fg^{-1} to 28.94 Fg^{-1} at 50 mVs^{-1}.

Another study discussed the influence of Ag incorporation to TiO_2 nanofibers on H_2S detecting characteristics. Pristine and Ag nanoparticles/TiO_2 (Ag/TiO_2) nanofibers were fabricated by Ma et al. [104]. H_2S sensing measurements showed that Ag/TiO_2 nanofibers have much higher detection responses and sensitivity than that of pristine Ag/TiO_2 nanofibers. It is found that the implementation of AgNPs leads to phase conversion of TiO_2 to rutile from anatase phase.

For the catalytic performance of TiO_2 nanofibers, Hao et al. [105] reported that mesoporous Au/TiO_2 nanofibers were obtained by incorporating Au nanoparticles (AuNPs) to TiO_2 nanofibers. The nanofibers were prepared via electrospinning a mixture solution of $Ti(OC_4H_9)_4$, $HAuCl_4$, and PVP. The effect of AuNPs loading (2, 5, and 10 wt%) toward the reduction of 4-nitrophenol (4-NP) to 4-aminophenol (4-AP) by sodium borohydride was reported. The results reported a decrease in the TOF value of the catalyst and an increase in the reaction rate constant with increasing AuNPs loading and this decline may be attributed to the small size and high surface area in case of 2 wt%, AuNPs. In addition, Tolba et al. [106] discussed the synthesis of ZnO nanobranches attaching TiO_2 nanofibers as nonprecious electrocatalyst for ethanol oxidation. The study reported the use of a colloidal solution of titanium isopropoxide, poly (vinyl acetate), and ZnNPs to produce electrospun fibers. Produced nanofibers went through a calcination and hydrothermal treatment. The study concluded that the concentration of ZnNPs has a distinct impact on the electrical conductivity and the electrocatalytic performance of the final nanofibers. For instance, utilizing 0.1 g ZnNPs in the electrospinnable solution produce active and stable nanofibers for ethanol oxidation in alkaline medium. In addition, Li et al. [107] measured the photocatalytic H_2-production activity in water splitting of 1D mesoporous NiO/TiO_2 composite nanofibers. The composite nanofibers were fabricated by electrospinning process of colloidal solution of tetra-n-butyl titanate (TBT), PVP, and $Ni(NO_3)_2.6H_2O$, and then the samples were calcined in air at 500°C for 3 h. It was reported that the H_2-production activity of TiO_2 nanofibers was developed at low NiO content (0–0.5 wt%). In fact, at 0.25 wt% of NiO, the mesoporous composite nanofibers exhibit the highest H_2-production activity. This can be related to the behavior of NiO as an active cocatalyst, which hinders the recombination of photogenerated charge carriers and diminishes the overpotential of H_2 production.

In another research for solar cells, Zr is doped into TiO_2 nanofibers to be utilized as photo-anodes of dye-sensitized solar cells (DSCs) [108]. A solution of PVAc, titanium isopropoxy, and zirconium n-propoxy was used to produce nanocomposite electrospun fibers. It was mentioned that the conversion efficiency of DSCs improved after Zr addition. The group tested the effect of Zr-doping amount (0.5, 1,1.5, and 2%) on TiO_2 nanofibers formulations for DSCs. The study reported that the photovoltaic efficiency reached 4.51% in case of 1% Zr-doping, while the efficiency was achieved to 1.61% for cells based on pristine TiO_2 nanofibers. This improvement can be attributed to the electrons transfer and dye loading. In more details, the presence of Zr developed the band gap through decreasing the charge recombination rate and, consequently, increases the electrons transfer (3.217 eV for 1% Zr-doped TiO_2 nanofibers and 3.202 eV for pristine TiO_2 nanofibers). Moreover, the dye loading of pristine TiO_2 nanofibers is found to be 0.52 $\times 10^{-7}$ molcm^{-2}, which is much lower than that of 1% Zr-doped TiO_2 (1.36 $\times 10^{-7}$ molcm^{-2}).

3.1.4. *Nickel oxide (NiO) nanofibers*

NiO nanofibers exhibit good chemical and thermal stability properties and can be prepared easily using electrospinning. High interest in preparing 3D nickel oxide (NiO) nanofibers technology to be applied in photovoltaic devices has been reported. This is because 3D NiO surfaces exhibit high degree of porosity, which increases the mobility of charge carriers and consequently enhances the efficiency of photovoltaic procedures [109]. Moreover, NiO has been used in batteries and chemical sensors industry because of its attractive characteristics and morphological properties [110, 111].

Macdonald et al. [112] prepared NiO nanofibers for cathode fabrication using electrospinning process of PAN and $Ni(AcAc)_2$. NiO nanofibers were deposited into fluorine-doped tin oxide (FTO) to produce p-type nano photocathodes with a surface area of 0.8 cm^2. To remove all organic residues, the electrode was sintered for 30 min at 450°C. The NiO nanofibers diameters were decreased to approximately 100 nm due to the calcination influence at 500°C and the nanofibers became more rough and straw-like structure owing to the elimination of PAN chains. Furthermore, Jian et al. [113] discussed the electrochemical performance of La-doped NiO nanofibers as positive electrode structures. La-doped NiO nanofibers are produced by the calcination procedure at 650°C for 3 h of electrospinning solution of $Ni(CH_3COO)_2.4H_2O$, $La(NO_3)_3.6H_2O$ and PVA. They designed asymmetric supercapacitor consisting of porous-activated carbon as a negative electrode, as-prepared La-doped nanofibers as a positive electrode, and 2M KOH aqueous solution as an electrolyte. It was seen that the perfect La/Ni loading ratio in the positive electrode was 1.5%. In addition, La-doping was found to develop the specific capacitance and the electrochemical stability. In more details, the specific capacitance became 94.85 F g^{-1}, which is 5.3 times better than the capacitors with undoped NiO nanofibers. Also, the coulomb efficiency of more than 90% is retained even after 1000 cycles.

NiO nanofibers exhibit excellent catalytic behavior. Elzatahry [114] described the use of PAN as a sacrificial template in the electrospinning method to fabricate NiO nanofibers to be designed as an electrocatalyst in methanol oxidation process in alkaline medium. The study reported the use of a solution of PAN and 50 wt% NiO to produce electrospun fibers. Uniform fibers with a diameter of 145–170 nm and low content of submicron and random nanofibers were produced. Porous NiO nanofibers are obtained after the elimination of PAN at 400°C for 3.5 h.

For sensing applications, Choi et al. [115] studied the fabrication of p-type NiO nanofibers and its activity for CO and NO_2 detection as a function of the size of the nanograins. The electrospinning solution consisted of 8.3 wt% PVA and 4 wt% nickel II acetate tetrahydrate. The produced nanofibers were subjected to calcination at 400–850°C to adjust the nanograins size. It was found that nanograins' size has a limited impact on the CO and NO_2 detection in case of p-type NiO nanofibers. On the other hand, the responses of n-type NiO nanofibers were highly dependent on the nanograins' size. Furthermore, Luo et al. [116] addressed the synthesis of Li-doped NiO nanofibers for nonenzymatic glucose sensing. The group reported the use of solution mixture of PVP, $Ni(NO_3)_2.6H_2O$, and $LiNO_3$ and another solution without $LiNO_3$, then calcined the product at 500°C for 3 h. It was noticed that the average diameter of the formed Li-doped NiO nanofibers was 200 nm, while that of NiO nanofibers was 150 nm. The fabricated Li-doped NiO nanofibers exhibit high surface area, high conductivity, and high electrocatalytic performance through holes and defective sites, which develop the electrons transfer process. In addition, SnO_2/NiO composite nanofibers were used in humidity sensing applications [117]. The study reported the use of electrospinning solution of $SnCl_2.2H_2O$, $NiCl_2.6H_2O$, and PVP and then annealed the produced fibers at 600°C for 3 h. It was noted that the high sensitivity of SnO_2/NiO nanofibers to humidity is ascribed to the active surface area that increases the conduction in the presence of water vapor.

3.2. Morphological characterization of metal oxide nanofibers

Morphological measurements of metal oxide nanofibers including fibers orientation, fibers distribution, and fibers diameters can be measured using many analytical instruments such as scanning electron microscope (SEM), X-ray diffraction (XRD), and X-ray photoelectron spectroscopy (XPS).

3.2.1. Scanning electron microscope characterization (SEM)

Scanning electron microscope (SEM), field emission scanning electron microscope (FESEM), transmission electron microscope (TEM), polarized light microscope, and atomic force microscope (AFM) can be utilized to study the nanofibers' diameters. It is worth mentioning that the high-resolution capability offered by SEM makes it a conventional technique to analyze nanofibers of which their structures are measured by their functionalities. The SEM is also used to study the chemical composition and surface topology and morphology. The nanograph's resolution is controlled by the interaction between the electron probes with metal oxide nanofibers. In more details, SEM measurements require conductive polymers and a small sample for morphology study. SEM analysis is a quick method used to measure the fiber diameters using a gold coating.

Figure 3 represents the SEM micrograph of electrospun ZnO nanofibers fabricated by Kanjwal et al. [95]. **Figures 3(A)** and **(B)** confirms the formation of continuous and smooth nanofibers of ZnAc/PVA. Furthermore, the addition of Zn NPs did not affect the structure of the produced nanofibers without any beads formation as clearly represented in **Figures 3(C)** and **(D)**. It should be also mentioned that the average diameters of the fabricated nanofibers are decreased after calcination process at 500°C for 90 min and this significant decrease can be attributed to the elimination of polymer matrix due to calcination at high temperatures. The shape of the nanofibers is not influenced by the addition of Zn NPs in general but there is an influence on the nanofibers' surface. Moreover, the surface of the nanofibers was relatively smooth in the case of Zn/Ac, while the addition of Zn NPs produced nanofibers with a rough surface.

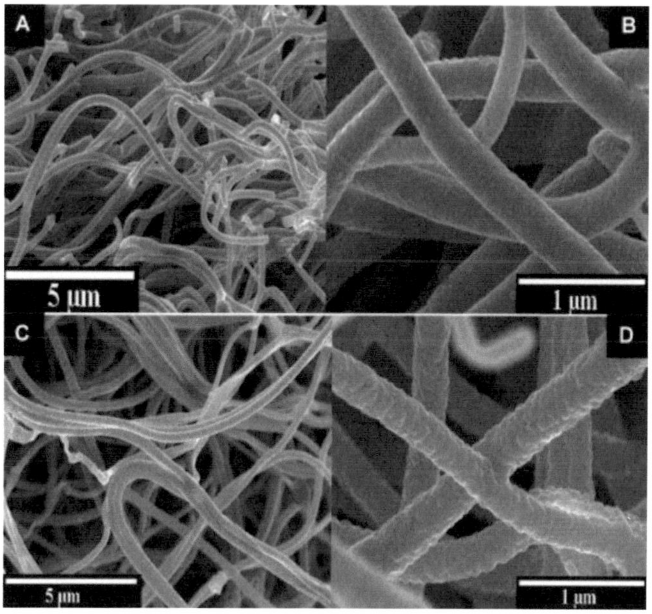

Figure 3. Low and high magnification SEM images of the powder obtained after calcination of ZnAc/PVA nanofibers mats (A and B), and Zn NPs/ZnAc/PVA (C and D) at 500°C for 90 min [95].

3.2.2. X-ray diffraction (XRD) characterization

XRD analysis is handled to study the crystal morphologies of various solids and metal fibers, including the lattice constants and geometry, the orientation of the single crystal, crystallite size, and crystal defects. The degree of crystallinity of the produced metal oxide nanofibers can be calculated from X-ray diffraction (XRD), both wide angle and small angle (WAXS and SAXS) and also from differential scanning calorimetry (DSC). A solid sample of metal oxide nanofibers is irradiated with X-rays of fixed wavelength while the angle of the diffracted X-ray beam is controlled as a function of the intensity of the diffracted beam. Moreover, each lattice plane satisfies Bragg's equation from which the values of the d-spacing between platelet sheets can be determined. XRD analysis was achieved to measure the crystal structures of the calcined ZnO nanofibers prepared by Kanjwal et al. [95] as described in **Figure 4**.

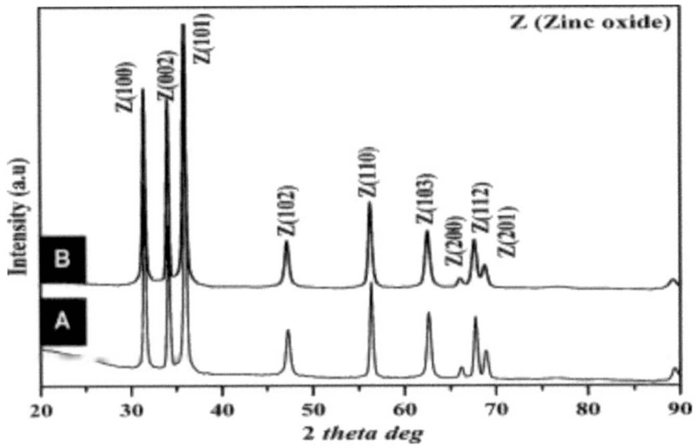

Figure 4. XRD patterns of nanofibers after calcination of (A) Zn NPs-free and (B) the hydrothermal product [95].

XRD results, in this case, confirm the synthesis of ZnO nanofibers. A SPECTRUM A of the annealed ZnO nanofibers confirms the formation of pure ZnO powder. It should be also noticed that the XRD spectra are not influenced by the incorporation of Zn NPs. In addition, the hydrothermal product has the same beaks as represented in spectra B.

3.2.3. X-ray photoelectron spectroscopy (XPS)

Another important characterization tool to investigate the surface chemistry of electrospun metal oxide nanofibers is X-ray photoelectron spectroscopy (XPS). The surface chemical nature

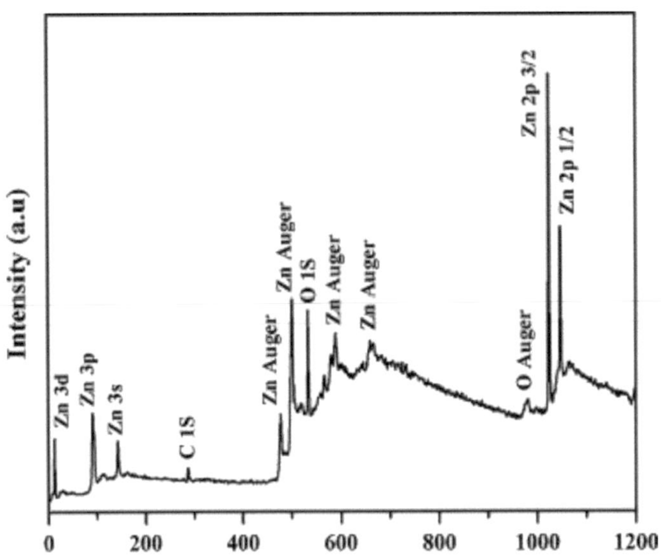

Figure 5. XPS data of the hydrothermally fabricated ZnO nanofibers [95].

of the metal oxide nanofibers can be estimated by its hydrophilicity which can be characterized by using XPS. XPS is performed to support the obtained XRD results and to measure the oxidation states and the changes in binding energies. The possible changes in binding energies of the prepared ZnO nanofibers have been reported by Kanjwal et al. [95] as represented in **Figure 5**. The peak of C 1s at 284 eV is related to the graphite tape employed in the sampling. The Zn 2p region in ZnO sample consists of the main 2p3/2 and 2p1/2 spin-orbit components with binding energies of 1020 and 1043 eV, respectively.

4. Conclusions

Electrospinning is an important technique for fabricating polymer and 1D metal oxide nanofibers. The desired characteristics can be measured by nanofibers' structure, which is controlled by various parameters including electrospinning solution, process, and ambient parameters. Selected polymers are critically used to prepare electrospun fibers' templates for this approach. The polymers used in this approach can be categorized into three classes including natural, synthetic, and copolymers. Electrospinning technique represents a vital promise with some borders for many applications such as small pore size and lack of cellular infiltration, which can be enhanced through multilayering and using of polymers of altered degradation grades. The technique has been adapted to produce many 1D metal oxide nanostructures for different important applications. This chapter introduced the fabrication of various metal oxide nanofibers which has attracted a significant attention because of their vital applications in the recent years such as tin oxide, zinc oxide, titanium oxide, and nickel oxide. Specific polymers were chosen as a sacrificial template to synthesize metal oxide nanofibers. After the annealing process, the organic specimens will be decomposed and the metal oxide nanofibers will be produced. Scanning electron microscope, X-ray diffraction, and X-ray photoelectron spectroscopy are used to characterize electrospun metal oxide nanofibers. The obtained data confirm the formation of homogeneously distributed metal oxide nanofibers.

Author details

Moustafa M. Zagho[1,2] and Ahmed Elzatahry[2*]

*Address all correspondence to: aelzatahry@qu.edu.qa

1 Center for Advanced Materials (CAM), Qatar University, Doha, Qatar

2 College of Arts and Science, Qatar University, Doha, Qatar

References

[1] Reneker DH, Yarin AL, Fong H, Koombhongse S. Bending instability of electrically charged liquid jets of polymer solutions in electrospinning. J Appl Phys. 2000;87:4531–4547.

[2] Schreuder-Gibson HL, Gibson P, Senecal K, Sennett M, Walker J, Yeomans W. Protective textile materials based on electrospun nanofibers. J Adv Mater. 2002;34:44–55.

[3] Huang ZM, Zhang YZ, Kotaki M, Ramakrishna S. A review on polymer nanofibers by electrospinning and their applications in nanocomposites. Compos Sci Technol. 2003;63:2223–2253.

[4] Theron SA, Yarin AL, Zussman E, Kroll E. Multiple jets in electrospinning: experiment and modeling. Polymer. 2005;46:2889–2899.

[5] Ma Z, Kotaki M, Inai R, Ramakrishna S. Potential of nanofiber matrix as tissue engineering scaffolds. Tissue Eng. 2005;11:101–109.

[6] Ahn YC, Park SK, Kim GT, Hwang YJ, Lee CG, Shin HS. Development of high efficiency nanofilters made of nanofibers. Curr Appl Phys. 2006;6:1030–1035.

[7] Lannutti J, Reneker D, Ma T, Tomasko D, Farson D. Electrospinning for tissue engineering scaffolds. Mater Sci Eng C. 2007;27:504–509.

[8] Hunley MT, Long TE. Electrospinning functional nanoscalefibers: a perspective for the future. Polym Int. 2008;57:385–389.

[9] Reneker DH, Yarin AL. Electrospinning jets and polymer nanofibers. Polymer. 2008;49:2387–2425.

[10] Zussman E, Theron A, Yarin AL. Formation of nanofiber crossbars in electrospinning. Appl Phys Lett. 2003;82:973–975.

[11] He J, Wan YQ, Yu JY. Scaling law in electrospinning: relationship between electric current and solution flow rate. Polymer. 2005;46:2799–2801.

[12] Luu YK, Kim K, Hsiao BS, Chu B, Hadjiargyrou M. Development of a nanostructured DNA delivery scaffold via electrospinning of PLGA and PLA-PEG block copolymers. J Control Release. 2003;89:341–353.

[13] Subbiah T, Bhat GS, Tock RW, Parameswaran S, Ramkumar SS. Electrospinning of nanofibers. J Appl Polym Sci. 2005;96:557–569.

[14] Ramakrishna S, Fujihara K, Teo WE, Yong T, Ma Z, Ramaseshan R. Electrospun nanofibers: solving global issues. Mater Today. 2006;9:40–50.

[15] Cui W, Zhou S, Li X, Weng J. Drug-loaded biodegradable polymeric nanofibers prepared by electrospinning. Tissue Eng. 2006;12:1070–1070.

[16] Wu Y, He JH, Xu L, Yu JY. Electrospinning drug-loaded poly (butylenes succinate-cobytylene terephthalate) (PBST) with acetylsalicylic acid (aspirin). Int J Electrospun Nanofibers Appl. 2007;1:1–6.

[17] Barnes CP, Sell SA, Knapp DC, Walpoth BH, Brand DD, Bowlin GL. Preliminary investigation of electrospun collagen and polydioxanone for vascular tissue engineering applications. Int J Electrospun Nanofibers Appl. 2007;1:73–87.

[18] Welle A, Kroger M, Doring M, Niederer K, Pindel E, Chronakis S. Electrospun aliphatic polycarbonates as tailored tissue scaffold materials. Biomaterials. 2007;28:2211–2219.

[19] Liang D, Hsiao BS, Chu B. Functional electrospun nanofibrous scaffolds for biomedical applications. Adv Drug Deliv Rev. 2007;59:1392–1412.

[20] Chong EJ, Phan TT, Lim IJ, Zhang YZ, Bay BH, Ramakrishna S. Evaluation of electrospun PCL/gelatinnanofibrous scaffold for wound healing and layered dermal reconstitution. Acta Mater. 2007;3:321–330.

[21] Ma ZW, Kotaki M, Ramakrishna S. Electrospun cellulose nanofiber as affinity membrane. J Membr Sci. 2005;265:115–123.

[22] Stankus JJ, Guan J, Wagner WR. Fabrication of biodegradable elastomeric scaffolds with sub-micron morphologies. J Biomed Mater Res. 2004;70A:603–614.

[23] Matthews JA, Wnek GE, Simpson DG, Bowlin GL. Electrospinning of collagen nanofibers. Biomacromolecules. 2002;3:232–238.

[24] Um IC, Fang DF, Hsiao BS, Okamoto A, Chu B. Electro-spinning and electro-blowing of hyaluronic acid. Biomacromolecules. 2004;5:1428–1436.

[25] Yarin AL, Zussman E. Upward needleless electrospinning of multiple nanofibers. Polymer. 2004;45:2977–2980.

[26] Tomaszewski W, Szadkowski M. Investigation of electrospinning with the use of a multi-jet electrospinning head. Fibres Text East Eur. 2005;13:22–26.

[27] Dosunmu OO, Chase GG, Kantaphinan W, Reneker DH. Electrospinning of polymer nanofibres from multiple jets on a porous tubular surface. Nanotechnology. 2006;17:1123–1127.

[28] Eichhorn SJ, Sampson WW. Statistical geometry of pores and statistics of porous nanofibrous assemblies. J R Soc Interface. 2005;2:309–318.

[29] Ekaputra AK, Prestwich GD, Cool SM, Hutmacher DW. Combining electrospun scaffolds with electrosprayed hydrogels leads to three-dimensional cellularization of hybrid constructs. Biomacromolecules. 2008;9:2097–2103.

[30] Cai SS, Liu YC, Shu XZ, Prestwich GD. Injectable glycosaminoglycan hydrogels for controlled release of human basic fibroblast growth factor. Biomaterials. 2005;26:6054–6067.

[31] Chew SY, Wen Y, Dzenis Y, Leong KW. The role of electrospinning in the emerging field of nanomedicine. Curr Pharm Des. 2006;12:4751-4770.

[32] Sill TJ, von Recum HA. Electrospinning: applications in drug delivery and tissue engineering. Biomaterials. 2008;29:1989–2006.

[33] Taylor GI. Electrically driven jets. Proc R Soc Lond A Math Phys Sci. 1969;313:453–475.

[34] Adomaviciute E, Rimvydas M. The influence of applied voltage on poly (vinyl alcohol) (PVA) nanofibre diameter. Fibers Text East Eur. 2007;15:64–65.

[35] Jiang HL, Fang DF, Hsiao BS, Chu B, Chen WL. Optimization and characterization of dextran membranes prepared by electrospinning. Biomacromolecules. 2004;5:326–333.

[36] Li M, Mondrinos MJ, Gandhi MR, Ko FK, Weiss AS, Lelkes PI. Electrospun protein fibers as matrices for tissue engineering. Biomaterials. 2005;26:5999–6008.

[37] Li J, He A, Zheng J, Han CC. Gelatin and gelatin-hyaluronic acid nanofibrous membranes produced by electrospinning of their aqueous solutions. Biomacromolecules. 2006;7:2243–2247.

[38] Li C, Vepari C, Jin HJ, Kim HJ, Kaplan DL. Electrospun silk-BMP-2 scaffolds for bone tissue engineering. Biomaterials. 2006;27:3115–3124.

[39] Zhang Y, Ouyang H, Lim CT, Ramakrishna S, Huang ZM. Electrospinning of gelatin fibers and gelatin/PCL composite fibrous scaffolds. J Biomed Mater Res B Appl Biomater. 2005;72:156–165.

[40] Zhang YZ, Venugopal J, Huang ZM, Lim CT, Ramakrishna S. Crosslinking of the electrospun gelatin nanofibers. Polymer. 2006;47:2911–2917.

[41] Zhong S, Teo WE, Zhu X, Beuerman RW, Ramakrishna S, Yung LYL. An aligned nanofibrous collagen scaffold by electrospinning and its effects on in vitro fibroblast culture. J Biomed Mater Res A. 2006;79A:456–463.

[42] Zeugolis DI, Khew ST, Yew ESY, Ekaputra AK, Tong YW, Yung LYL. Electro-spinning of pure collagen nano-fibres—just an expensive way to make gelatin? Biomaterials. 2008;29:2293–2305.

[43] Kwon IK, Matsuda T. Co-electrospun nanofiber fabrics of poly (L-lactide-co-ε-caprolactone) with type I collagen or heparin. Biomacromolecules. 2005;6:2096–2105.

[44] Yang L, Fitie CFC, Werf KOV, Bennink ML, Dijkstra PJ, Feijen J. Mechanical properties of single electrospun collagen type I fibers. Biomaterials. 2008;29:955–962.

[45] Neal RA, McClugage III SG, Link MC, Sefcik LS, Ogle RC, Botchwey EA. Laminin nanofiber meshes that mimic morphological properties and bioactivity of basement membranes. Tissue Eng Part C. 2008;15:11–21.

[46] Hakkarainen M. Aliphatic polyesters: abiotic and biotic degradation and degradation products. Adv Polym Sci. 2002;157:113–138.

[47] Jing Z, Xu XY, Chen XS, Liang QZ, Bian XC, Yang LX. Biodegradable electrospun fibers for drug delivery. J Control Release. 2003;92:227–231.

[48] Boland ED, Wnek GE, Simpson DG, Pawlowski KJ, Bowlin GL. Tailoring tissue engineering scaffolds using electrostatic processing techniques: a study of poly (glycolic acid) electrospinning. J Macromol Sci Pure Appl Chem A. 2001;38:1231–1243.

[49] Kenawy ER, Bowlin GL, Mansfield K, Layman J, Simpson DG, Sanders EH. Release of tetracycline hydrochloride from electrospun poly (ethylene-co-vinylacetate, poly (lactic acid), and a blend. J Control Release. 2002;81:57–64.

[50] Ho BC, Lee YD, Chin WK. Thermal degradation of polymethacrylic acid. J Polym Sci Polym Chem. 1992;30:2389–2397.

[51] Bhattarai N, Cha DI, Bhattarai SR, Khil MS, Kim HY. Biodegradable electrospun mat: novel block copolymer of poly (p-dioxanone-co-lactide)-block-poly (ethylene glycol). J Polym Sci B Polym Phys. 2003;41:1955–1964.

[52] Dharmaraj N, Kim CH, Kim KW, Kim HY, Suh EK. Spectral studies of SnO_2 nanofibers prepared by electrospinning method. Spectrochim Acta A. 2006;64:136–140.

[53] Guan H, Shao C, Chen B, Gong J, Yang X. A novel method for making CuO superfine fibers via an electrospinning technique. Inorg Chem Commun. 2003;6:1409–1411.

[54] Macias M, Chacko A, Ferraris JP, Balkus KJ. Electrospunmesoporous metal oxide fibers. Micropor Mesopor Mater. 2005;86:1–13.

[55] Yang XH, Shao CL, Guan HY, Li XL, Gong J. Preparation and characterization of ZnO nanofibers by using electrospun PVA/zinc acetate composite fiber as precursor. Inorg Chem Commun. 2004;7:176–178.

[56] Onozuka K, Ding B, Tsuge Y, Naka T, Yamazaki M, Sugi S, Ohno S, Yoshikawa M, Shiratori S. Electrospinning processed nanofibrous TiO_2 membranes for photovoltaic applications. Nanotechnology. 2006;17:1026–1031.

[57] Yang XH, Shao CL, Liu YC, Mu RX, Guan HY. Nanofibers of CeO_2 via an electrospinning technique. Thin Solid Films. 2005;478:228–231.

[58] Wang D, Chu XF, Gong ML. Gas-sensing properties of sensors based on single-crystalline SnO_2 nanorods prepared by a simple molten-salt method. Sens Actuators B Chem. 2006;117:183–187.

[59] Yang MR, Chu SY, Chang RC. Synthesis and study of the SnO_2 nanowires growth. Sens Actuators B Chem. 2007;122:269–273.

[60] Luo SH, Fan JY, Liu WL, Zhang M, Song ZT, Lin CL, Wu XL, Chu PK. Synthesis and low-temperature photoluminescence properties of SnO_2 nanowires and nanobelts. Nanotechnology. 2006;17:1695–1699.

[61] Comini E, Faglia G, Sberveglieri G, Calestani D, Zanotti L, Zha M. Tin oxide nanobelts electrical and sensing properties. Sens Actuators B Chem. 2005;111:2–6.

[62] Hu JQ, Bando Y, Golberg D. Self-catalyst growth and optical properties of novel SnO_2 fishbone-like nanoribbons. Chem Phys Lett. 2003;372:758–762.

[63] Ding B, Wang M, Yu J, Sun G. Gas sensors based on electrospun nanofibers. Sensors. 2009;9:1609–1624.

[64] Ding B, Kim J, Miyazaki Y, Shiratori S. Electrospun nanofibrous membrane coated quartz crystal microbalance as gas sensor for NH_3 detection. Sens Actuators B Chem. 2004;101:373–380.

[65] Liu HQ, Kameoka J, Czaplewski DA, Craighead HG. Polymeric nanowire chemical sensor. Nano Lett. 2004;4:671–675.

[66] Kessick R, Tepper G. Electrospun polymer composite fiber arrays for the detection and identification of volatile organic compounds. Sens Actuators B Chem. 2006;117:205–210.

[67] Luoh R, Hahn HT. Electrospun nanocomposite fiber mats as gas sensors. Compos Sci Technol. 2006;66:2436–2441.

[68] Kim ID, Rothschild A, Lee BH, Kim DY, Jo SM, Tuller HL. Ultrasensitive chemiresistors based on electrospun TiO_2 nanofibers. Nano Lett. 2006;6:2009–2013.

[69] Zhang Y, He XL, Li JP, Miao ZJ, Huang F. Fabrication and ethanol-sensing properties of micro gas sensor based on electrospun SnO_2 nanofibers. Sens Actuators B Chem. 2008;132:67–73.

[70] Wang G, Ji Y, Huang XR, Yang XQ, Gouma PI, Dudley M. Fabrication and characterization of polycrystalline WO_3 nanofibers and their application for ammonia sensing. J Phys Chem B. 2006;110:23777–23782.

[71] Khorami HA, Keyanpour-Rad M, Vaezi MR. Synthesis of SnO_2/ZnO composite nanofibers by electrospinning method and study of its ethanol sensing properties. Appl Surf Sci. 2011;257:7988–7992.

[72] Zhao Y, He X, Li J, Cao X, Jia J. Porous CuO/SnO_2 composite nanofibers fabricated by electrospinning and their H_2S sensing properties. Sens Actuators B Chem. 2012;165:82–87.

[73] Park JY, Asokan K, Choi SW, Kim S. Growth kinetics of nanograins in SnO_2 fibers and size dependent sensing properties. Sens Actuators B Chem. 2011;152:254–260.

[74] Zhang Y, Li J, An G, He X. Highly porous SnO_2 fibers by electrospinning and oxygen plasma etching and its ethanol-sensing properties. Sens Actuators B Chem. 2010;144:43–48.

[75] Santos JP, Fernández MJ, Fontecha JL, Matatagui D, Sayago I, Horrillo MC, Gracia I. Nanocrystalline tin oxide nanofibers deposited by a novel focused electrospinning

method. Application to the detection of TATP precursors. Sensors. 2014;14:24231–24243.

[76] Cheng L, Ma SY, Wang TT, Li XB, Luo J, Li WQ, Mao YZ, GZ DJ. Synthesis and characterization of SnO_2 hollow nanofibers by electrospinning for ethanol sensing properties. Mater Lett. 2014;131:23–26.

[77] Choi JK, Hwang IS, Kim SJ, Park JS, Park SS, Jeong U, Kang YC, Lee JH. Design of selective gas sensors using electrospun Pd-doped SnO_2 hollow nanofibers. Sens Actuators B. 2010;150:191–199.

[78] Li WQ, Ma SY, Li YF, Li XB, Wang CY, Yang XH, Cheng L, Mao YZ, Luo J, Gengzang DJ, Wan GX, Xu XL. Preparation of Pr-doped SnO_2 hollow nanofibers by electrospinning method and their gas sensing properties. J Alloys Compd. 2014;605:80–88.

[79] Du H, Wang J, Su M, Yao P, Zheng Y, Yu N. Formaldehyde gas sensor based on SnO_2/In_2O_3 hetero-nanofibers by a modified double jets electrospinning process. Sens Actuators B. 2012;166:746–752.

[80] Tang W, Wang J, Yao P, Li X. Hollow hierarchical SnO_2-ZnO composite nanofibers with heterostructure based on electrospinning method for detecting methanol. Sens Actuators B. 2014;192:543–549.

[81] Hiroyuki U. Influence of surfactant micelles on morphology and photoluminescence of zinc oxide nanorods prepared by one-step chemical synthesis in aqueous solution. J Phys Chem C. 2007;111:9060–9065.

[82] Kind H, Yan HQ, Messer B, Law M, Yang PD. Nanowire ultraviolet photodetectors and optical switches. Adv Mater. 2002;14:158–160.

[83] Izaki M, Watase S, Takahashi H. Room-temperature ultraviolet light-emitting zinc oxide micropatterns prepared by low-temperature electrodeposition and photoresist. App Phys Lett. 2003;83:4930–4932.

[84] Hingorani S, Pillai V, Kumar P, Multani MS, Shah DO. Microemulsion mediated synthesis of zinc-oxide nanoparticles for varistor studies. Mater Res Bull. 1993;28:1303–1310.

[85] Sakohara S, Ishida M, Anderson MA. Visible luminescence and surface properties of nanosized ZnO colloids. J Phys Chem B. 1998;102:10169–10175.

[86] Zhang Z, Liu S, Chow S, Han MY. Modulation of the morphology of ZnO nanostructures via aminolytic reaction: from nanorods to nanosquamas. Langmuir. 2006;22:6335–6340.

[87] Ochanda F, Cho K, Andala D, Keane TC, Atkinson A, Jones WE. Synthesis and optical properties of co-doped ZnO submicrometer tubes from electrospun fiber templates. Langmuir. 2009;25:7547–7552.

[88] Huang MH, Feick H, Weber E, Wu Y, Tran N, Yang P. Catalytic growth of zinc oxide nanowires by vapor transport. Adv Mater. 2001;13:113–116.

[89] Kong XY, Wang ZL. Polar-surface dominated ZnO nanobelts and the electrostatic energy induced nanohelixes, nanosprings and nanospirals. Appl Phys Lett. 2004;84:975–978.

[90] Hughes WL, Wang ZL. Formation of piezoelectric single-crystal nanorings and nanobows. J Am Chem Soc. 2004;126:6703–6709.

[91] Wang ZL, Kong XY, Zuo JM. Induced growth of asymmetric nanocantilever arrays on polar surfaces. Phys Rev Lett. 2003;91:185502–185505.

[92] Ding B, Ogawa T, Kim J, Fujimoto K, Shiratori S. Fabrication of a superhydrophobic nanofibrous zinc oxide film surface by electrospinning. Thin Solid Films. 2008;516:2495–2501.

[93] Barakat NAM, Abadir MF, Sheikh FA, Kanjwal MA, Park SJ, Kim HY. Polymeric nanofibers containing solid nanoparticles prepared by electrospinning and their applications. Chem Eng J. 2010;156:487–495.

[94] Zhang Z, Li X, Wang C, Wei L, Liu Y, Shao C. ZnO hollow nanofibers: fabrication from facile single capillary electrospinning and applications in gas sensors. J Phys Chem C. 2009;113:19397–19403.

[95] Kanjwal MA, Sheikh FA, Barakat NAM, Li X, Kim HY, Chronakis IS. Zinc oxide's hierarchical nanostructure and its photocatalytic properties. Appl Surf Sci. 2012;258:3695–3702.

[96] Mauro AD, Zimbone M, Fragalà ME, Impellizzeri G. Synthesis of ZnO nanofibers by the electrospinning process. Mater Sci Semicond Process. 2016;42:98–101.

[97] Li W, Ma S, Yang G, Mao Y, Luo J, Cheng L, Gengzang D, Xu X, Yan S. Preparation, characterization and gas sensing properties of pure and Ce doped ZnO hollow nanofibers. Mater Lett. 2015;138:188–191.

[98] Park M, Han SM. Enhancement in conductivity through Ga, Al dual doping of ZnO nanofibers. Thin Solid Films. 2015;590:307–310.

[99] Wu MS, Wang MJ, Jow JJ, Yang WD, Hsieh CY, Tsai HM. Electrochemical fabrication of anatase TiO_2 nanostructure as an anode material for aqueous lithium-ion batteries. J Power Sources. 2008;185:1420–1424.

[100] Bojinov V, Grabchevb I. Synthesis and properties of new adducts of 2,2,6,6-tetramethylpiperidine and 2-hydroxyphenylbenzotriazole as polymer photostabilizers. J Photochem Photobiol A Chem. 2002;150:223–231.

[101] Manickam M, Singh P, Issa TB, Thurgate S. Electrochemical behavior of anatase TiO_2 in aqueous lithium hydroxide electrolyte. J Appl Electrochem. 2006;36:599–602.

[102] Reiman KH, Brace KM, Smith TJG, Nandhakumar I, Attard GS, Owen JR. Lithium insertion into TiO_2 from aqueous solution—facilitated by nanostructure. Electrochem Commun. 2006;8:517–522.

[103] He X, Yang CP, Zhang GL, Shi DW, Huang QA, Xiao HB, Liu Y, Xiong R. Supercapacitor of TiO_2 nanofibers by electrospinning and KOH treatment. J Mater Des. 2016;106:74–80.

[104] Ma S, Jia J, Tian Y, Cao L, Shi S, Li X, Wang X. Improved H_2S sensing properties of Ag/TiO_2 nanofibers. Ceram Int. 2016;42:2041–2044.

[105] Hao Y, Shao X, Li B, Hu L, Wang T. Mesoporous TiO_2 nanofibers with controllable Au loadings for catalytic reduction of 4-nitrophenol. Mater Sci Semicond Process. 2015;40:621–663.

[106] Tolba GMK, Barakat NAM, Bastaweesy AM, Ashour EA, Abdelmoez W, El-Newehy MH, Al-Deyab SS, Kim HY. Hierarchical TiO_2/ZnO nanostructure as novel non-precious electrocatalyst for ethanol electrooxidation. J Mater Sci Technol. 2015;31:97–105.

[107] Li L, Cheng B, Wang Y, Yu J. Enhanced photocatalytic H_2-production activity of bicomponent NiO/TiO_2 composite nanofibers. J Colloid Interface Sci. 2015;449:115–121.

[108] Mohamed IMA, Dao VD, Barakat NAM, Yasin AS, Yousef A, Choi HS. Efficiency enhancement of dye-sensitized solar cells by use of ZrO_2-doped TiO_2 nanofibers photoanode. J Colloid Interface Sci. 2016;476:9–19.

[109] Liu M, Wang Y, Li P, Cheng Z, Zhang Y, Zhang M, Hu M, Li J. Preparation and characterization of multilayer NiOnano-products via electrospinning. Appl Surf Sci. 2013;284:453–458.

[110] Qu Y, Zhou W, Miao X, Li Y, Jiang L, Pan K, Tian G, Ren Z, Wang G, Fu H. A new layered photocathode with porous NiO nanosheets: an effective candidate for p-type dye-sensitized solar cells. Chemistry. 2013;8:3085–3090.

[111] Sialvi MZ, Mortimer RJ, Wilcox GD, Teridi AM, Varley TS, Wijayantha KGU, Kirk CA. Electrochromic and colorimetric properties of Nickel(II) oxide thin films prepared by aerosol-assisted chemical vapor deposition. ACS Appl Mater Interfaces. 2013;5:5675–5682.

[112] Macdonald TJ, Xu J, Elmas S, Mange YJ, Skinner WM, Xu H, Nann T. NiO nanofibers as a candidate for a nanophotocathode. Nanomaterials. 2014;4:256–266.

[113] Jian J, Luo F, Gao C, Suo C, Wang X, Song H, Hu X. Synthesis of La-doped NiO nanofibers and their electrochemical properties as electrode for supercapacitors. Ceram Int. 2014;40:6973–6977.

[114] Elzatahry A. Polyacrylonitrile electrospun nanofiber as a template to prepare NiO nanostructure electrocatalyst. Int J Electrochem Sci. 2014;9:22–31.

[115] Choi JM, Byun JH, Kim SS. Influence of grain size on gas-sensing properties of chemiresistive p-type NiO nanofibers. Sens Actuators B. 2016;227:149–156.

[116] Luo X, Zhang Z, Wan Q, Wu K, Yang N. Lithium-doped NiO nanofibers for non-enzymatic glucose sensing. Electrochem Commun. 2015;61:89–92.

[117] Pascariu P, Airinei A, Olaru N, Petrila I, Nica V, Sacarescu L, Tudorache F. Microstructure, electrical and humidity sensor properties of electrospun NiO-SnO_2 nanofibers. Sens Actuators B. 2016;222:1024–1031.

Chapter 2

Electrospinning Functional Polyacrylonitrile Nanofibers with Polyaniline, Carbon Nanotubes, and Silver Nitrate as Additives

Nuray Kizildag and Nuray Ucar

Additional information is available at the end of the chapter

http://dx.doi.org/10.5772/65472

Abstract

Electrospinning of composite nanofibers has been attracting great attention as a way of producing functional nanofibers. Composite nanofibers are produced with the incorporation of the additives into the polymer melt/solution before electrospinning process and reported to show many superior properties such as high modulus, increased strength, improved thermal stability, or some new functionalities such as flame retardancy, antimicrobial properties, water repellency, soil resistance, decreased gas permeability, electromagnetic shielding, electrical conductivity, and so on. The availability of the wide range of additives makes it possible to produce a wide range of functional nanocomposite nanofibers that are promising for various applications. Polyaniline (PANI) as an inherently conductive polymer is being investigated as an additive for improving conductivity. Carbon nanotubes (CNT) are widely used for either their reinforcement ability or their superior electrical conductivity. Silver nanoparticles (AgNPs) are being incorporated into polymer matrices to obtain antibacterial activity. This chapter provides a comprehensive review about polyacrylonitrile (PAN) nanofibers with PANI, CNTs, AgNPs, and their combinations and highlights the synergistic effects obtained by their combined use.

Keywords: antibacterial, antistatic, semiconductive, multifunctional, nanocomposite, electrospinning, nanofiber, polyacrylonitrile, polyaniline, carbon nanotubes, silver nitrate

1. Introduction

Electrospinning is a process that uses electrostatic forces to produce nanofibers. The setup for electrospinning consists of a high-voltage power supply that is used to charge the electrospinning solution, a pump that is used to feed the solution through the needle and a grounded collector that is used to collect the nanofibers in the nanoweb form. In the electric field forming between the needle and the collector, a jet occurs when the electrostatic force overcomes the surface tension of the solution droplet and it undergoes bending and whipping instability as a result of which the solvent evaporates and nanofibers form [1, 2]. Electrospun nanofibers have attracted great attention due to their unique properties, ease of fabrication, and possibilities of functionalization [1]. They have high surface-area-to-volume ratio, low density, and high pore volume [3–5], which qualify them for a number of applications such as tissue engineering [6], wound healing [7], drug delivery [8], filtration [9], sensors [10], energy harvesting and storage [11], polymer reinforcement [12], and so on.

While nanofibers have already been providing unpaired properties, composite nanofibers made the way one step further and realized additional functionalities, which showed the potential of improving the applications of nanofibers [13]. The composite nanofibers are produced with the incorporation of some additives. It has been possible to produce semiconductive nanofibers with the addition of PANI [14], antibacterial nanofibers with the addition of AgNPs [15], and semiconductive nanofibers with improved mechanical properties with the addition of CNTs [16]. Furthermore, some synergistic effects have been reported to occur when the additives are used together.

This chapter provides a comprehensive review of the studies about electrospun composite polyacrylonitrile nanowebs produced using PANI, CNTs, and AgNPs as additives; highlights the synergistic effects obtained by their combined use; and shows that it is possible to produce multifunctional nanocomposite nanowebs with the combined use of the additives.

2. Polyacrylonitrile nanowebs with polyaniline (PAN/PANI)

PANI is one of the most widely investigated conductive polymer due to its environmental stability, low cost of raw material, ease of synthesis, and good compatibility with polymer supports, controllable electrical conductivity and interesting redox properties associated with its nitrogen chain [17–19]. Insolubility of PANI has been a big problem retarding its applications. In this regards, the counter-ion-induced solubility of PANI has been a successful attempt that made polyanilines in conducting form to be soluble in some ordinary organic solvents [20]. Since then, solution blending of PANI with polystyrene [21], polyimide [22], polyamide [23], polyacrylonitrile [24], so on has been reported. The method of solution blending has also been adapted to prepare electrospinning solutions of different polymers with PANI with the aim of producing conductive nanofibers which may be used in sensor [25], tissue engineering [26], supercapacitor [27], flexible solar cells [28], hydrogen storage applications [29], and so on.

PANI exists in various levels of oxidation such as fully reduced leucoemeraldine base form, the half-oxidized emeraldine base form (EB), and the fully oxidized pernigraniline base form [30, 31]. The EB oxidation state of PANI can be doped with a protonic acid to form emeraldine salt (ES), which transforms the electronic structure of the chain into a polaronic lattice and results in an electrically conductive state [30]. For doping, nonvolatile acid dopants, such as camphorsulfonic acid (CSA) and dodecylbenzylsulfonic acid (DBSA), are widely used since they overcome the evaporation disadvantage of smaller molecular organic and inorganic acids, which cause conductivity depression of the acid-doped PANI [32]. The properties of the dopant, such as molecular weight, molecular size, acidity, and so on determine the doping ability of the dopant [33, 34] and thus the conductivity of the solutions, which has a direct influence on the electrospinning process and the properties of the nanowebs. Solvent selection is also important since it primarily determines the solubility of the polymers and electrospinnability [35]. The studies about PAN/PANI composite nanofibers are mainly focused on the effects of PANI content, dopant types, solvent types, solution preparation procedures, and redoping on the properties of composite nanowebs [14, 19, 32, 35–37].

Raeesi et al. [19] investigated PAN/PANI nanowebs with different composition ratios using N-methyl-2-pyrrolidone (NMP) as the solvent. Horizontal electrospinning setup with stationary collector was used for nanofiber production with various contents of PANI (up to 30 wt%) at various electrospinning temperatures. While bead-on-string structure with nonuniform morphology was observed at the PANI content of 30%, drops instead of fibers were observed at the PANI content above 30%. Average nanofiber diameter decreased with increasing PANI content. The electrical conductivity of the nanowebs was measured as 10^{-1} S/cm for nanofibers with 30% PANI after a doping process with HCl vapor [19]. Qavamnia et al. [36] electrospun PAN nanofibers with CSA-doped PANI using NMP as the solvent and investigated the effect of PANI content on the morphological, electrical conductivity, and mechanical properties of the blend nanofibers. PANI content was varied as 0, 10, 20, 30, and 40 wt%. Composite nanofiber diameters were in the range of 59 to 234 nm and decrease was observed in nanofiber diameter with the increase in PANI content. Beads were observed at 40 wt% PANI content. Tenfold increase in tensile modulus and 3-fold increase in tenacity were observed at the PANI content of 30 wt%. The electrical conductivity was reported to be between 10^{-10} S/cm and 10^{-1} S/cm depending on the amount of PANI added to nanofiber structure [36]. Kizildag et al. [14] electrospun nanofibers of PAN- and CSA-doped PANI using dimethyl sulfoxide (DMSO) as the solvent and investigated the effect of PANI content and the application of different dissolution methods on the morphology, chemical structure, conductivity, crystallinity, mechanical, and thermal properties of nanowebs. PANI content was varied as 1, 3, 5, 7, 10, and 30 wt%. For the investigation of the effect of dissolution process, three samples with 10 wt% PANI were prepared using different preparation procedures. Two different PANI solutions, which were stirred magnetically for 2 and 10 days, and another solution, which was exposed to homogenization with an ultrasonic probe for 1.5 h after magnetic stirring, were prepared. The diameters of the composite nanofibers increased until the PANI content of 5% and then decreased as the PANI content was further increased. The composite nanofibers were generally uniform except the nanofibers with 30 wt% PANI which had nonhomogeneous fiber structure. PAN/PANI composite nanowebs with 1, 5, 10 wt% PANI appeared to have improved crystallinity values in comparison to neat PAN nanofibers. Breaking stress decreased with PANI addition, while breaking elongation increased as PANI content increased until the content of 7 wt% and then decreased. The conductivity of the composite nanofibers was improved, reaching a value higher than 10^{-6} S/cm with 3 wt% PANI which was in the range for electrostatic discharge applications. Thermal stability of the nanofibers was improved with PANI addition. Increase in dissolution time and application of ultrasonic homogenization affected the diameter, mechanical properties, crystallinity, and thermal properties of the nanofibers, while they had negligible effects on conductivity [14]. Kizildag et al. [35] also investigated the effects of different dopants such as camphorsulfonic acid (CSA), dodecylbenzene sulfonic acid (DBSA) in isopropanol, and dodecylbenzene sulfonic acid sodium salt ($DBSANa^+$), and different solvents such as NMP and *N,N*-dimethylformamide (DMF) on the structure and properties of PAN/PANI composite nanowebs. It was shown that the solvents and dopants had significant effects on morphology, average nanofiber diameter, mechanical properties, while they had less effect on conductivity of the composite nanofiber webs. The fibers produced from NMP solvent generally had larger fiber diameters than the fibers produced from DMF, while the use of $DBSANa^+$ resulted in the formation of larger diameters in comparison to other dopants. The use of NMP as the solvent resulted in higher breaking stress values for the reference samples and the composite samples, which contained CSA-doped PANI. The conductivity values of the composite nano/microfiber webs were around 10^{-8} and 10^{-9} S/cm [35]. In another study, comparing the effects of the solvents such as DMF and DMSO, and the dopants such as CSA and DBSA in isopropanol and $DBSANa^+$, Kizildag et al. resulted that the composite nanofibers of PAN/PANI produced from DMSO generally had larger fiber diameters than nanofibers produced from DMF. While the diameter of composite nanofibers with the dopants CSA and DBSA (in isopropanol) decreased compared to 100% PAN nanofiber, the diameter of nanofiber with $DBSANa^+$drastically increased. Increased breaking elongations were observed for the samples electrospun from DMSO, while decrease was observed for the samples electrospun from DMF. CSA used with DMSO resulted in the highest conductivity of 10^{-6} S/cm, which is in the range suitable for electrostatic discharge applications [37]. Ucar et al. [32]. investigated the effect of different solvents (DMSO, NMP, and DMF) and solvent mixtures, application of dispersion and mixing techniques during solution preparation and redoping process on PAN/PANI composite nanofibers using CSA as the dopant. The morphology, average nanofiber diameters, crystallinity, mechanical properties, thermal properties, and electrical conductivity were all affected by the solvents used. Mechanical dispersion technique resulted in higher tensile breaking stress values than the corresponding magnetic stirring. While redoping did not affect the morphology and the diameter of the nanofibers significantly, it affected the tensile properties of the nanowebs by increasing the breaking stress values and decreasing the elongation values. The conductivity was improved 10 times and reached 1.2×10^{-5} S/cm after redoping [32].

As seen from the literature, polyaniline addition affected the morphology, nanofiber diameter, crystallinity, mechanical properties, electrical properties, and thermal properties of the nanofibers [14, 19, 32, 35–37].

3. Polyacrylonitrile nanowebs with carbon nanotubes (PAN/CNTs)

CNTs are widely used as additives in nanofiber production for either reinforcement or functionalization. They possess special properties such as high strength and aspect ratio, good thermal and electrical conductivities, and a low density, which are all important in the preparation of polymer composites [38–40]. They can have diameters ranging from 1 to 100 nm and lengths of up to millimeters. Their densities can be as low as 1.3 g/cm^3 and their Young's moduli are greater than 1 TPa. The weakest types of CNTs have strengths of several GPa [41]. Their electrical conductivity can be as high as 10^6 S/m [42]. In most cases, the addition of only a few percentages of CNTs to the nanofiber structure results in enhanced tensile properties, thermal stability, electrical properties and dimensional stability [38–43]. Nevertheless, to be able to fully benefit from their reinforcing properties, a uniform dispersion, and orientation of CNTs in the polymer matrix are important [16]. The strength and elongation are adversely affected by the addition of CNTs especially when the CNTs are not well dispersed and aligned along the fiber axis. The agglomerates can act as stress points instead of reinforcing and result in a decrease in both tensile strength and breaking elongation [43]. For better dispersion, several approaches, which can be roughly classified as mechanical and chemical methods, can be seen in literature. While the mechanical methods such as ball milling, ultrasonication, and high shear mixing contribute to better dispersion by altering the surface energy of the solids, chemical methods such as surface functionalization improve the chemical compatibility between the CNTs and the solvents and the polymer matrixes, enhance wetting characteristics and reduce their tendency to agglomerate [16]. Ultrasonication is the most common method applied for dispersion of the CNTs [16, 44]. CNTs are ultrasonicated in either the solvent or the polymer solution. In many studies, CNTs are chemically treated before ultrasonication [40, 45, 46]. Besides all the efforts to better disperse the CNTs in polymers, electrospinning is reported to be a process that greatly contributes to the alignment and dispersion of CNTs by charge, confinement, and flow effects [47–49]. Dror et al. established a model to explain how the CNT-polymer composite nanofibers were formed by electrospinning. The randomly oriented MWCNT rods in the electrospinning solution were oriented along the streamlines of the electrospinning solution due to elongation of the fluid jet [49]. In addition, it has been shown that significant interactions exist between PAN chains and CNTs, which lead to better dispersion of CNTs in PAN. DMF, which is widely used as a solvent for PAN provides another advantage. It is a good solvent for suspending oxidized CNTs [35].

Ge et al. [50] prepared PAN/CNT nanofibers on an electrospinning setup with a rotating collector using acid treated CNTs. The CNT content was changed as 3, 5, 1, 20 wt%. UV-visible spectroscopy indicated that there was a strong interfacial bonding between the CNTs and PAN macromolecules. The orientation of the CNTs within the nanofibers was observed to be much higher than that of the PAN polymer crystal matrix. Incorporation of CNTs has been demonstrated to enhance electrical conductivity, tensile modulus, thermal deformation temperature, and decomposition temperature of composite nanowebs. At the CNT content of 20 wt%, the electrical conductivity the composite nanofibers was measured as 1.0 S/cm [50]. Hou et al. [45] produced composite nanofibers of PAN with oxidized MWNTs. While the surfaces of the pure PAN nanofibers and composite nanofibers with low amount of

CNTs were smooth, they became rough with the increase in CNT content. Increase was observed in tensile modulus and tensile strength, while decrease was observed in breaking elongation with CNT addition. The tensile modulus reached 4.4 GPa at 20 wt% of MWCNT with a 144% improvement. The maximum tensile strength was 80.0 MPa at about 5% MWCNT with a 75% improvement. It was shown that a higher concentration of MWCNTs effectively resisted heat shrinkage of the composite nanowebs during carbonization [45]. Heikkilä and Harlin [44] electrospun pure, salt-containing, and CNT-containing nanofibers using different nozzle sizes, spinning voltages and distances. PAN and additive concentration were selected as 13 and 0.25 wt%, respectively. Composite nanofibers with CNTs showed more pronounced surface roughness and markedly larger fiber diameters than pure PAN and salt-containing nanofibers. The electrospinnability was improved with the addition of CNTs. They resulted that the solution composition had a greater effect on nanofiber diameter than process parameters [44]. Saeed et al. [46] used 1 and 2 wt% MWNTs functionalized by Friedel-Crafts acylation to produce PAN/CNT nanofibers. Functionalization provided better dispersion of the CNTs, which resulted in nanofibers with less beads and higher mechanical properties. Specific tensile strength and the specific modulus increased while breaking elongation decreased at CNT content of 1 wt%. Increase was observed in the degradation temperature with CNT addition [46]. Chen et al. [40] functionalized CNTs by grafting with PAN through the process of plasma-induced grafting polymerization before incorporating into PAN nanofibers. PAN grafted CNTs provided stable and well-dispersed solutions due to the chemical affinity between the polar-modified groups and the organic solvent. TEM observations showed the CNTs were generally parallel and oriented along the axes of the nanofibers. The surfaces of the composite nanofibers became rough with the increase in CNT content. Raman results indicated enhanced growth of graphitic crystals in the carbonized PAN due to the presence of the CNTs. The sheet resistance of the CNT/carbon nonwoven fabrics was appreciably enhanced by increasing CNT concentration. Carbonized PAN/CNT nanowebs had a significant SE of more than 30 dB at 30 MHz, and the SE was around 10–15 dB between 900 MHz to 3.0 GHz, even for the small thickness of 150 μm. The results indicated that carbonized PAN/CNT nanowebs were promising for use as effective and practical EMI shielding materials due to their lightweight, good mechanical properties, low cost, and high shielding performance [40]. Yousefzadeh et al. [47] dispersed MWNTs in DMF using probe sonicator for about 1 h. Magnetic stirrer was used to mix the polymer with CNT dispersion until the polymer was uniformly dissolved in the solvent. The CNT content was varied as 0.01, 0.05, 0.1, 0.3, 0.5, 1, and 2 wt%. According to the SEM images of dispersed nanotubes, mixing for about 1 h with 30% amplitude was found to be sufficient to achieve a well-dispersed solution. To improve the dispersion of nanotubes at high concentration, MWNTs were refluxed in HNO_3 and stirred at 15°C for 8 h to attach functional groups of carboxyl and hydroxyl groups onto MWNTs. As the CNT content increased, the surfaces of the composite nanofibers became rough. They obtained thicker fibers with the addition MWNTs compared to CNT-free ones. The highest tensile strength, tensile modulus, and breaking elongation were obtained with 1% CNT addition [47]. Qiao et al. [51] functionalized the SWNTs by polymer wrapping, dispersed them in DMF through mild bath sonication for 2 h, added PAN, and mechanically stirred overnight at 40°C using a magnetic stirrer to yield a homogeneous solution. The SWNT content was varied as 0.25, 0.5, 0.75, and 1 wt%. While the surfaces of the nanofibers became rough, the diameters of the nanofibers became larger with CNT addition. The introduction of SWNTs improved the modulus and tensile strength of the PAN nanowebs. The tensile strength of the nanocomposites at about 0.75 wt% SWNTs was increased by 58.9%. In addition, the tensile modulus showed a peak value of 4.62 GPa with 66.8% improvement. While the electrical conductivity increased to 2.5 S/cm, T_g increased by about 3°C by incorporating 0.75 wt% SWNTs into the PAN matrix [51]. Wang et al. [52] functionalized MWCNTs by Friedel–Crafts acylation. They used *in situ* and *ex situ* solution polymerization systems to obtain composite solutions with 0.5, 1, 3, and 5 wt% CNTs and an electrospinning setup with a stationary collector to produce composite nanofibers. While the tensile modulus, tensile strength, thermal stability, crystallinity were improved with CNT addition regardless of the incorporation method, the *in situ* polymerization resulted in better dispersion, finer fiber formation, higher orientation and crystallinity, larger crystal size, higher tensile modulus and tensile strength, higher thermal stability [52]. Almuhamed et al. [16] dispersed MWNTs (0.2, 0.4, 0.5, 0.7, 1.0, and 1.5 wt%) in the solvent by high shear mixing using a homogenizer device and ultrasonication, added PAN to the dispersions and stirred magnetically until dissolvation of PAN. While PAN nanofibers had a mean diameter of 568 nm, the composite nanofibers had average diameters ranging from 325 to 795 depending on the percentage of MWNTs. The composite nanofibers possessed an electrical volume percolation threshold at very low loading percentage of MWNTs corresponding to 0.5 wt%. While the volumetric electrical conductivity increased by five orders of magnitude from 1.85×10^{-11} S/m at CNT content of 0.4 wt% to 4.15×10^{-6} S/m at 0.5 wt%, the surface electrical conductivity was not very much affected by CNT addition. The composite nanowebs were suggested for use as pressure sensors as an exponential relationship was observed between the mechanical pressure applied and the volume conductivity [16]. Eren et al. [53] investigated the effects of differently functionalized (carboxyl, amine, and hydroxyl functionalized) MWNTs on the structure and properties of composite PAN nanofibers produced by horizontal electrospinning setup with a rotating collector. CNTs were dispersed in DMF by ultrasonic tip followed by the application of ultrasonic bath for 45 min. The concentration of PAN and CNTs were 7 and 1 wt%, respectively. Slight increases were observed in nanofiber diameters with CNT addition regardless of the functionalization process applied. Among the three differently functionalized MWNTs, amine-functionalized nanotubes provided the highest tensile strength, tensile modulus, and crystallinity. The conductivities of the composite nanowebs were measured as around 10^{-7} S/cm regardless of the functional groups of CNTs [53].

According to the literature, significant differences are reported to occur in properties of PAN/CNT nanowebs depending on the types of CNTs used, functionalization processes applied to the CNTs, solution preparation methods, additive contents, and electrospinning conditions. The studies show that improvement in electrospinnability, crystallization, mechanical properties, thermal properties, thermal stability, and electrical conductivity can be obtained with the addition of CNTs if the optimum conditions are ensured.

4. Polyacrylonitrile nanowebs with silver nanoparticles (PAN/AgNPs)

Silver has been the most widely used material to fight against broad range of microorganisms since ancient times [54]. AgNPs are expected to show better performance than microparticles due to their increased surface area. Moreover, they have been found to exhibit remarkable catalytic activity, and high electrical conductivity [55]. The incorporation of AgNPs into nanofibers is reported to result in an improved mechanical properties and desirable functionalities, such as antistatic and antibacterial properties [56], which offer great potential in various fields such as filtration, protective textiles, medical textiles, biomedical applications, and so on [4]. The studies in literature have demonstrated the antibacterial activity of nanowebs containing AgNPs against *Staphylococcus aureus, Escherichia coli, Bacillus Subtilis, Bacillus cereus, and Pseudomonas aeruginosa* [4, 15, 57, 58]. For antimicrobial performance of the nanofibers, the rate and the amount of silver release are important factors [4, 57]. A steady and prolonged release of silver cations can inhibit the growth of bacteria when their concentration is above 0.1 ppb [59].

Figure 1. Photographs of PAN/10 wt% $AgNO_3$ solution: (a) without any treatment; (b) after exposure to xenon-arc lamp for 15 min; (c) after exposure to xenon-arc lamp for 30 min.

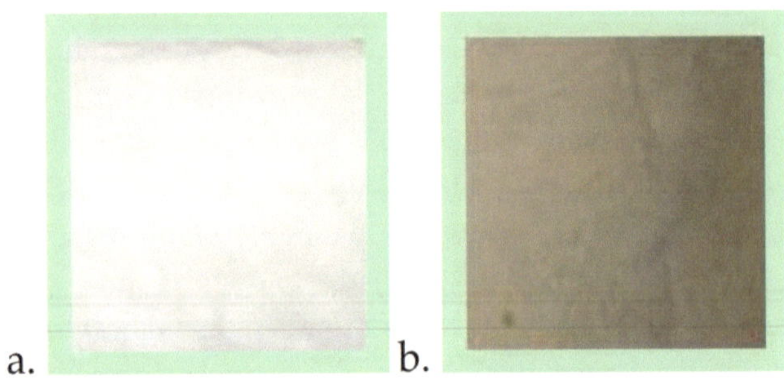

Figure 2. Photographs of PAN/10 wt% $AgNO_3$ nanoweb: (a) before; (b) after chemical reduction process performed by immersing the nanowebs into aqueous solution of hydrazinium hydroxide.

There are two general approaches in the preparation of polymeric nanowebs with AgNPs such as *ex situ* and *in situ*. In the *ex situ* methods, AgNPs are synthesized before their incorporation into the polymer that is used as the dispersion medium; whereas in *in situ* methods, the polymer is used as the reaction medium and AgNPs are synthesized in the polymer. Providing better dispersion of AgNPs, *in situ* methods have attracted more attention than *ex situ* methods in preparing polymeric matrix with AgNPs [60, 61]. Chemical reduction by the solvent [15], or aqueous solutions of sodium borohydride ($NaBH_4$) [15], and hydrazinium hydroxide (N_2H_5OH) [62], photo reduction [57, 62] and heat treatment [15] are some of the methods used for *in situ* synthesis of AgNPs [15, 57]. With the formation of AgNPs, electrospinning solutions and nanowebs change color from yellow to brown [63]. The photographs showing the color changing in the solutions and nanowebs after reduction process are presented in **Figures 1** and **2**, respectively.

PAN [56], cellulose acetate [64], chitosan [65], gelatin [60], polyvinyl alcohol [66], and so on have been used in nanofiber formation as host polymers for *in situ* formation of AgNPs. PAN is an excellent polymer for *in situ* synthesis of AgNPs since coordination bonds forming between silver ions and cyano nitrogen of PAN avoid aggregation of AgNPs [55].

Lee et al. [67] prepared 7 wt% PAN/DMF solution with silver nitrate (0.05, 0.2, and 0.5 wt% of the amount of PAN) and aged the solutions for 10 days before electrospinning, thereby using DMF as the reducing medium. AgNPs were spherical in shape and had an average diameter of 5.8 nm. UV-Visible spectra showed that the amount of AgNPs increased in time during aging without any change in their sizes. When the concentration of PAN was increased, the generation of AgNPs were slower. They concluded that DMF successfully reduced the Ag^+ ions and PAN acted as a stabilizing agent to inhibit the agglomeration of the Ag nanoparticles. Decrease in nanofiber diameter was observed with increase in AgNP amount which was attributed to the increased conductivity of the electrospinning solutions [67]. Wang et al. [68] prepared composite PAN nanofibers with $AgNO_3$ and applied chemical reduction in N_2H_5OH aqueous solution. AgNPs with average diameters of 10 nm were dispersed homogeneously in PAN nanofibers. D and G peaks in Raman spectrum of the composite nanowebs indicated that the structure of PAN became similar to the PAN-based carbon fiber after being doped with AgNPs. This was attributed to the AgNPs acting as a catalyst for the dehydrogenation of hydrocarbon compound at room temperature [68]. Sichani et al. [57] added 0.05, 0.2, and 0.5 wt% $AgNO_3$ to PAN/DMF solutions and used xenon arc lamp in order to *in situ* synthesize AgNPs in these solutions. The average size of silver nanoparticles was around 10 nm. Improvement in electrospinning process, decrease in the number of the beads and fiber diameter were observed with the addition of up to 0.5 wt% $AgNO_3$. The crystallinity increased with the addition of $AgNO_3$. The composite nanofibers had steady and good antibacterial activity, especially against *P. aeruginosa*. The antibacterial activity against *S. aureus* and *E. coli* was lower [57]. Rujitanaroj et al. [63] prepared 10 w/v% PAN/DMF solutions with different contents of $AgNO_3$ (0.5, 1.5, 2.5 wt%) and aged the solutions for 5 days before electrospinning. The composite nanowebs were exposed to UV-irradiation to enhance the reduction process. UV treatment resulted in the formation of larger AgNPs on the surface of nanofibers. The diameter of the nanofibers decreased dramatically with increase in the $AgNO_3$ amount and decreased further slightly with the UV treatment. The amount of AgNPs increased with aging time without any change in their size. While the tensile strength increased slightly, elongation at break decreased slightly with $AgNO_3$ addition. The cumulative amounts of the silver release increased as the submersion time, the initial $AgNO_3$ concentration and the UV irradiation time interval increased. The antibacterial activity of the membranes against *S. aureus* and *E. coli* bacteria increased with the increase in $AgNO_3$ concentration and UV irradiation time [63]. Lee et al. [69] prepared composite PAN nanofibers with $AgNO_3$ and performed chemical reduction in N_2H_5OH aqueous solution. AgNPs with diameters of less than 5.8 nm were dispersed homogeneously in PAN nanofibers and the nanofibers were effective against *S. aureus* and *E. coli* [69]. Shi et al. [4] added $AgNO_3$ (0.5 and 1.25 wt%) to 8 wt% PAN/DMF solution and performed atmospheric plasma treatment for the reduction of silver ions in the PAN/$AgNO_3$/DMF solutions. AgNPs with diameters ranging between 3 and 6 nm were found to be uniformly dispersed in the nanofiber. The composite nanofibers showed excellent antibacterial activity against both Gram-positive and Gram-negative microorganisms [4]. Mahapatra et al. [15] prepared composite nanowebs of PAN with 8 wt% $AgNO_3$ and applied three different reduction methods such as refluxing the solution before electrospinning, treating the composite nanoweb with aqueous solution of sodium borohydride ($NaBH_4$), and heating the prepared composite nanofibers at 160°C. Distinct absorption band at 410 nm on UV-Visible spectra confirmed the formation of AgNPs. TEM micrographs showed that the AgNPs were dispersed homogeneously on the surface of PAN nanofibers, and particle diameter was about 5–15 nm. They concluded that refluxing resulted in smaller nanofiber diameter than chemical reduction applied to the nanoweb and attributed this to the increased conductivity of the electrospinning solution during refluxing. All three methods resulted in good antibacterial activity against *S. aureus, E. coli,* and *B. subtilis*. The size of the silver nanoparticles were smallest and length of inhibition zone was the highest against *S. aureus* bacteria for the composite membrane that was prepared by refluxing Ag^+ ions in DMF solvent [15]. Ucar et al. [56] prepared PAN/AgNP nanofibers with and without stabilizer and applied four different reduction methods (refluxing in DMF, chemical reduction by hydrazinium hydroxide, xenon arc reduction in solution and nanoweb form) in order to compare their effects on the properties of the composite nanowebs. 10 wt% $AgNO_3$ was added to 10 wt% PAN/DMF solution. Horizontal electrospinning setup with a rotating collector was used. Higher absorbance values were obtained for PAN/10 wt% $AgNO_3$ nanowebs reduced with hydrazinium hydroxide. While the xenon arc reduction applied to the solution resulted in thicker nanofibers, chemical reduction process resulted in finer nanofibers. Composite nanowebs electrospun from the composite solution exposed to xenon-arc displayed the lowest breaking strength values. The nanowebs which were chemically treated with aqueous solution of N_2H_5OH displayed the highest breaking strength. Electrical conductivity was measured as 10^{-7} S/cm for the composite nanowebs reduced by chemical reduction and xenon arc application. With the addition of the stabilizer, fiber diameter increased while conductivity decreased [56]. Demirsoy et al. [62] investigated the effect of dispersion technique, reduction method, and AgNP amount on the properties of PAN/AgNP nanofibers. $AgNO_3$ (1 wt%, 3 wt%, and 10 wt%) was added to 7 wt % PAN/DMF solution and stirred for 1 h. The nanowebs were produced on a horizontal electrospinning setup with a rotating collector. Two different reduction methods such as chemical reduction by N_2H_5OH solution and xenon arc reduction were applied. While the reduction process by N_2H_5OH applied to the nanowebs resulted in a decrease, reduction by xenon arc applied to the electrospinning solution resulted in an increase in the nanofiber diameters. The crystallinity of PAN/10 wt% $AgNO_3$ nanoweb was lower than that of PAN/1 wt% $AgNO_3$ nanoweb. The highest tensile strength was obtained at the $AgNO_3$ content of 1

http://dx.doi.org/10.5772/65472

wt% after chemical reduction by N_2H_5OH. The tensile strength decreased with the increase in additive amount. Application of an additional dispersion process by ultrasonic homogenizer and bath generally provided higher breaking strength values than only mixing with a magnetic stirrer due to better dispersion of particles. The electrical conductivity of PAN/AgNP nanowebs was around 10^{-7} S/cm at 3 wt% $AgNO_3$ content and a decrease was observed with the increase in $AgNO_3$ content. Excellent antibacterial activity against *S. aureus* was obtained at the $AgNO_3$ content of 3 wt% after the chemical reduction process [62].

According to the existing literature on PAN/AgNP nanofibers, it can be concluded that concentration of the AgNPs, the dispersion and reduction methods applied, addition of a stabilizer are important factors that affect the final properties of the composite nanowebs.

5. Composite PAN nanowebs with the combined addition PANI, CNTs and AgNPs

A recent approach for functionalization of nanofibers is the combined use of additives. There are studies about composite electrospun nanowebs and films showing improved mechanical properties, thermal stability, crystallization and antimicrobial activity [70], electrical properties [71, 72], and biocompatibility [72] with the combined use of CNTs and AgNPs. Besides there are also some studies reporting about the synergistic effects obtained by the simultaneous use of the conductive polymers, CNTs and AgNPs [73, 74].

Ucar et al. [75] produced PAN composite nanofibers adding PANI and CNT simultaneously. While the diameters of nanofibers increased, the effect of PANI on diameter was higher than that of CNTs. The breaking of the composite nanofibers with 1% CNT and 3% PANI was 25% higher than that of pure PAN nanofibers. Conductivity of the composite nanowebs was in the semiconductive range regardless of the additive content. The crystallinity of PAN/PANI/CNT composite nanofiber was higher than that of pure PAN, PAN/CNT, and PAN/PANI composite nanofibers [75]. Eren et al. [76] incorporated various amounts of CNTs, AgNPs, and PANI into PAN nanofibers in order to see the synergistic effect of the additives on the final properties of the composite materials. Increase in the amount and types of additives generally resulted in an increase in the diameter of nanofibers and decrease in mechanical strength. Composite nanofibers with AgNPs displayed higher breaking strength and electrical conductivity than the composite nanofibers with CNTs. Generally, PANI improved the crystallinity of the composite nanowebs more than the nanoparticles. The use of the additives (PANI, CNT, AgNPs) at low concentration resulted in an increase in the temperature and enthalpy for cyclization compared to pure PAN nanofiber. Even though each of the nanoparticles was used in low concentrations, the composite nanowebs of PAN/1 wt% CNT/1 wt% $AgNO_3$ and PAN/ 3 wt% PANI/1 wt% $AgNO_3$ exhibited antimicrobial properties due to the synergistic effect of additives. It was suggested that PAN composite nanofibers with 3 wt% PANI and 1 wt% $AgNO_3$ generally presented better performance than the other samples in terms of electrical conductivity, antimicrobial activity, mechanical strength, crystallization, and thermal stability [76]. Kizildag et al [70] produced composite nanofibers from a solution of PAN, MWNTs, and $AgNO_3$ in DMSO by the electrospinning method. They immersed the composite nanowebs into aqueous solution of hydrazinium hydroxide for the chemical reduction of silver ions. PAN/f-MWNTs/AgNPs nanowebs displayed enhanced conductivity and antimicrobial properties particularly when the chemical reduction process was applied. Besides, they showed improved crystallinity. While the reduction process made the highest contribution to the ultimate tensile strength, elongation, and conductivity of the nanowebs, MWNT content had negligible effect on conductivity of the nanowebs. PAN with 1 wt% MWNTs and 1 wt %$AgNO_3$ was suggested for use as antistatic and antibacterial nanowebs [70].

6. Conclusions

The composite nanofibers have been attracting great interest, as they display many improved properties such as high modulus, increased strength, improved thermal stability, electrical, barrier properties, and/or new functionalities such as flame retardancy, antimicrobial properties, water repellency, conductivity, and so on. Polyaniline, carbon nanotubes, and silver nanoparticles are widely used additives in the production of composite nanofibers. While their incorporation into the nanofiber structure is reported to affect morphological properties, chemical structure, crystallinity, conductivity, thermal properties, mechanical properties, and so on, polyaniline is mainly added to improve the conductivity; carbon nanotubes for improving strength, conductivity and thermal properties; and silver nanoparticles for developing antibacterial properties. The studies on the combined use of these additives are promising since it has been possible to obtain some synergistic effects as well as multifunctionality. With the improvements in the processing that will ensure especially uniform dispersion of the additives and higher production rates, the potential applications of functional composite nanofibers will soon turn into reality.

Author details

Nuray Kizildag* and Nuray Ucar

*Address all correspondence to: kizildagn@itu.edu.tr

Textile Engineering Department, Istanbul Technical University, Istanbul, Turkey

References

[1] Fang J, Wang X, Lin T. Functional applications of electrospun nanofibers. In: Lin T, editor. Nanofibers-production, properties and functional applications. Crotia:Intech; 2011. p. 287–327. DOI: 10.5772/916.

[2] Picciani PHS, Medeiros ES, Pan Z, Orts WJ, Mattoso LHC, Soares BG. Development of conducting polyaniline/poly(lactic acid) nanofibers by electrospinning. Journal of Applied Polymer Science. 2009;112:744–753. DOI: 10.1002/app.29447.

[3] Shi X, Zhou W, Ma D, Ma Q, Bridges D, Ma Y, Hu A. Electrospinning of nanofibers and their applications for energy devices. Journal of Nanomaterials. 2015;2015(140716):1–20. DOI: 10.1155/2015/140716.

[4] Shi Q, Vitchuli N, Nowak J, Caldwell JM, Breidt F, Bourham M, Zhang X, McCord M. Durable antibacterial Ag/polyacrylonitrile (Ag/PAN) hybrid nanofibers prepared by atmospheric plasma treatment and electrospinning. European Polymer Journal. 2011;47(7):1402–1409. DOI: 10.1016/j.eurpolymj.2011.04.002.

[5] Reneker D, Yarin A. Electrospinning jets and polymer nanofibers. Polymer. 2008;49(10): 2387–2425. DOI: 10.1016/j.polymer.2008.02.002.

[6] Ito Y, Hasuda H, Kamitakahara M, Ohtsuki C, Tanihara M, Kang IK, Kwon OH. A composite of hydroxyapatite with electrospun biodegradable nanofibers as a tissue engineering material. Journal of Bioscience and Bioengineering. 2005;100(1):43–49. DOI: 10.1263/jbb.100.43.

[7] Khil MS, Cha DI, Kim HY, Kim IS, Bhattarai N. Electrospun nanofibrous polyurethane membrane as wound dressing. Journal of Biomedical Materials Research, Part B: Applied Biomaterials. 2003;67(2):675–679. DOI: 10.1002/jbm.b.10058.

[8] Yang D, Li Y, Nie J. Preparation of gelatin/PVA nanofibers and their potential application in controlled release of drugs. Carbohydrate Polymers. 2007;69(3):538–543. DOI: 10.1016/j.carbpol.2007.01.008.

[9] Barhate RS, Ramakrishna S. Nanofibrous filtering media: Filtration problems and solutions from tiny materials. Journal of Membrane Science. 2007;296(1–2):1–8. DOI: 10.1016/j.memsci.2007.03.038.

[10] Wang X, Drew C, Lee S, Senecal KJ, Kumar J, Samuelson LA. Electrospun nanofibrous membranes for highly sensitive optical sensors. Nano Letters. 2002;2(11):1273–1275. DOI: 10.1021/nl020216u.

[11] Wang L, Yu Y, Chen PC, Zhang DW, Chen CH. Electrospinning synthesis of C/Fe3O4 composite nanofibers and their application for high performance lithium ion batteries. Journal of Power Sources. 2008;183(2):717–723. DOI: 10.1016/j.jpowsour.2008.05.079.

[12] Bergshoef MM, Vancso GJ. Transparent nanocomposites with ultrathin, electrospun nylon-4,6 fiber reinforcement. Advanced Materials. 1999;11(16):1362–1365. DOI: 10.1002/(SICI)1521-4095(199911)11:16<1362::AID-ADMA1362>3.0.CO;2-X.

[13] Smole MS, Hribernik S, Veronovski N, Kurecic M, Kleinschek KS. Electrokinetic properties of nanocomposite fibres. In: Reddy B, editor. Advances in nanocomposites – synthesis, characterization and industrial applications. Crotia:Intech; 2011. p. 403–429. DOI: 10.5772/604.

[14] Kizildag N, Ucar N, Karacan I, Onen A, Demirsoy N. The effect of the dissolution process and the polyaniline content on the properties of polyacrylonitrile/polyaniline composite nanoweb. Journal of Industrial Textiles. 2016;45(6):1548–1570. DOI: 10.1177/1528083714564636.

[15] Mahapatra A, Garg N, Nayak BP, Mishra BG, Hota G. Studies on the synthesis of electrospun PAN-Ag composite nanofibers for antibacterial application. Journal of Applied Polymer Science. 2012;124:1178–1185. DOI: 10.1002/app.35076.

[16] Almuhamed S, Khenoussi N, Schacher L, Adolphe D, Balard H. Measuring of electrical properties of MWNT-reinforced PAN nanocomposites. Journal of Nanomaterials. 2012;2012(750698):1–7. DOI: 10.1155/2012/750698.

[17] Zhai G, Fan Q, Tang Y, Zhang Y, Pan D, Qin Z. Conductive composite films composed of polyaniline thin layers on microporous polyacrylonitrile surfaces. Thin Solid Films. 2010;519:169–173. DOI: 10.1016/j.tsf.2010.07.088.

[18] Pan W, Yang SL, Li G, Jiang JM. Electrical and structural analysis of conductive polyaniline/polyacrylonitrile composites. European Polymer Journal. 2005;41:2127–2133. DOI: 10.1016/j.eurpolymj.2005.04.003.

[19] Raeesi F, Nouri M, Haghi AK. Electrospinning of polyaniline-polyacrylonitrile blend nanofibers. e-Polymers. 2009;114:1–13. DOI: 10.1515/epoly.2009.9.1.1350.

[20] Cao Y, Smith P, Heeger AJ. Counter-ion induced processability of conducting polyaniline and of conducting polyblends of polyaniline in bulk polymers. Synthetic Metals. 1992;48:91–97. DOI: 10.1016/0379-6779(92)90053-L.

[21] Segal E, Haba Y, Narkis M. On the structure and electrical conductivity of polyaniline/polystyrene blends prepared by an aqueous-dispersion blending method. Journal of Polymer Science Part B: Polymer Physics. 2001;39(5):611–621. DOI: 10.1002/1099-0488(20010301)39:5<611::AID-POLB1035>3.0.CO;2-R.

[22] Liangcai L, Ming W, Huoming S. Preparation and EIS studies on polyimide/polyaniline blend film for corrosion protection. Polymers for Advanced Technologies. 2001;12(11–12):720–723. DOI: 10.1002/pat.94.

[23] Zhang Q, Wang X, Geng Y, Chen D, Jing X. Morphology and thermal properties of conductive polyaniline/polyamide composite films. Journal of Polymer Science: Part B: Polymer Physics. 2002;40:2531–2538. DOI: 10.1002/polb.10316.

[24] Pan W, He X, Chen Y. Preparation and characterization of polyacrylonitrile-polyaniline blend nanofibers. Applied Mechanics and Materials. 2010;44–47:2195–2198. DOI: 10.4028/www.scientific.net/AMM.44-47.2195.

[25] Bishop-Haynes A, Gouma P. Electrospun polyaniline composites for NO2 detection. Materials and Manufacturing Processes. 2007;22(6):764–767. DOI: 10.1080/10426910701385408.

[26] Li M, Guo Y, Wei Y, MacDiarmid AG, Lelkes PI. Electrospinning polyaniline-contained gelatin nanofibers for tissue engineering applications. Biomaterials. 2006;27(13):2705–2715. DOI: 10.1016/j.biomaterials.2005.11.037.

[27] Chaudhari S, Sharma Y, Archana PS, Jose R, Ramakrishna S, Mhaisalkar S, Srinivasan M. Electrospun Polyaniline nanofibers web electrodes for supercapacitors. Journal of Applied Polymer Science. 2013;129(4):1660–1668. DOI: 10.1002/APP.38859.

[28] Peng S, Zhu P, Wu Y, Mhaisalkara SG, Ramakrishna S. Electrospun conductive polyaniline–polylactic acid composite nanofibers as counter electrodes for rigid and flexible dye-sensitized solar cells. RSC Advances. 2012;2:652–657. DOI: 10.1039/C1RA00618E.

[29] Phani AR, Robin De Britto MT, Srinivasan S, Stefanakos L. Polyaniline nanofibers obtained by electrospin process for hydrogen storage applications. International Journal of Environmental Research and Development. 2014;4(4):375–386.

[30] Premvardhan L, Peteanu LA, Wang PC, MacDiarmid AG. Electronic properties of the conducting form of polyaniline from electroabsorption measurements. Synthetic Metals. 2001;116:157–161. DOI: 10.1016/S0379-6779(00)00477-X.

[31] Jin J, Wang Q, Haque M. Doping dependence of electrical and thermal conductivity of nanoscale polyaniline thin films. Journal of Physics D: Applied Physics. 2010;43(20): 205302. DOI: 10.1088/0022-3727/43/20/205302.

[32] Ucar N, Kizildag N, Onen A, Karacan I, Eren O. Polyacrylonitrile-polyaniline composite nanofiber webs: Effects of solvents, redoping process and dispersion technique. Fibers and Polymers. 2015;16(10):2223–2236. DOI: 10.1007/s12221-015-5426-3.

[33] Long Y, Chen Z, Wang N, Zhang Z, Wan M. Resistivity study of polyaniline doped with protonic acids. Physica B. 2003;325:208–213. DOI: 10.1016/S0921-4526(02)01526-0.

[34] Li G, Zheng P, Wang NL, Long YZ, Chen ZJ, Li JC, Wan MX. Optical study on doped polyaniline composite films. Journal of Physics: Condensed Matter. 2004;16:6195–6204. DOI: 10.1088/0953-8984/16/34/018.

[35] Kizildag N, Ucar N, Onen A, Karacan I. Polyacrylonitrile/polyaniline composite nano/microfiber webs produced by different dopants and solvents. Journal of Industrial Textiles. 2016;46(3):787–808. DOI: 10.1177/1528083715598654.

[36] Qavamnia SS, Nasouri K. Conductive polyacrylonitrile/polyaniline nanofibers prepared by electrospinning process. Polymer Science Series A. 2015;57(3):343–349. DOI: 10.1134/S0965545X1503013X.

[37] Kizildag N, Ucar N, Onen A, Karacan I. Polyacrylonitrile/polyaniline composite nanofiber webs with electrostatic discharge properties. Journal of Composite Materials. DOI: 10.1177/0021998316630583.

[38] Iijima S. Helical microtubules of graphitic carbon. Nature. 1991;354:56–58. DOI: 10.1038/354056a0.

[39] Baughman RH, Zakhidov AA, de Heer WA. Carbon nanotubes – the route toward applications. Science. 2002;297(5582):787–792. DOI: 10.1126/science.1060928.

[40] Chen IH, Wang CC, Chen CY. Fabrication and structural characterization of polyacrylonitrile and carbon nanofibers containing plasma-modified carbon nanotubes by electrospinning. The Journal of Physical Chemistry C. 2010;114:13532–13539. DOI: 10.1021/jp103993b.

[41] Coleman JN, Khan U, Blau WJ, Gun'ko YK. Small but strong: A review of the mechanical properties of carbon nanotube–polymer composites. Carbon. 2006;44:1624–1652. DOI: 10.1016/j.carbon.2006.02.038.

[42] Min BG, Chae HG, Minus ML, Kumar S. Polymer/carbon nanotube composite fibers – An overview, In: Lee KP, Gopalan AI, Marquis FDS, editors. Functional composites of carbon nanotubes and applications. Kerala, India: Transworld Research Network; 2009. p. 43–73.

[43] Ayutsede J, Gandhi M, Sukigara S, Ye H, Hsu CM, Gogotsi Y, Ko F. Carbon nanotube reinforced Bombyx mori silk nanofibers by the electrospinning process. Biomacromolecules. 2006;7:208–214. DOI: 10.1021/bm0505888.

[44] Heikkilä P, Harlin A. Electrospinning of polyacrylonitrile (PAN) solution: Effect of conductive additive and filler on the process. eXPRESS Polymer Letters. 2009;3(7):437–445. DOI: 10.3144/expresspolymlett.2009.53.

[45] Hou H, Ge JJ, Zeng J, Li Q, Reneker DH, Greiner A, Cheng SZD. Electrospun polyacrylonitrile nanofibers containing a high concentration of well-aligned multiwall carbon nanotubes. Chemistry of Materials. 2005;17:967–973. DOI: 10.1021/cm0484955.

[46] Saeed K, Park SY. Preparation and characterization of multiwalled carbon nanotubes/polyacrylonitrile nanofibers. Journal of Polymer Research. 2010;17:535–540. DOI: 10.1007/s10965-009-9341-4.

[47] Yousefzadeh M, Amani-Tehran M, Latifi, M, Ramakrishan S. Morphology and mechanical properties of polyacrylonitrile/multi-walled carbon nanotube (PAN/MWNTs) nanocomposite electrospun nanofibers. Transaction F: Nanotechnology. 2010;17(1):60–65.

[48] Ahn BW, Chi YS, Kang TJ. Preparation and characterization of multi-walled carbon nanotube/poly(ethylene terephthalate) nanoweb. Journal of Applied Polymer Science. 2008;110:4055–4063. DOI: 10.1002/app.28968.

[49] Dror Y, Salalha W, Khalfin RL, Cohen Y, Yarin AL, Zussman E. Carbon nanotubes embedded in oriented polymer nanofibers by electrospinning. Langmuir. 2003;19(17):7012–7020. DOI: 10.1021/la034234i.

[50] Ge JJ, Hou H, Li Q, Graham MJ, Greiner A, Reneker DH, Harris FW, Cheng SZD. Assembly of well-aligned multiwalled carbon nanotubes in confined polyacrylonitrile

environments: Electrospun composite nanofiber sheets. Journal of the American Chemical Society. 2004;126:15754–15761. DOI: 10.1021/ja048648p.

[51] Qiao B, Ding X, Hou X, Wu S. Study on the electrospun CNTs/polyacrylonitrile-based nanofiber composites. Journal of Nanomaterials. 2011;2011(839462):1–7. DOI: 10.1155/2011/839462.

[52] Wang K, Gu M, Wang J, Qin C, Dai L. Functionalized carbon nanotube/polyacrylonitrile composite nanofibers: Fabrication and properties. Polymers for Advanced Technologies. 2012;23:262–271. DOI: 10.1002/pat.1866.

[53] Eren O, Ucar N, Onen A, Karacan I, Kizildag N, Demirsoy N, Vurur OF, Borazan I. Effect of differently functionalized carbon nanotubes on the properties of composite nanofibers. Indian Journal of Fibre and Textile Research. 2016;41:138–144.

[54] Jung WK, Koo HC, Kim KW, Shin S, Kim SH, Park YH. Antibacterial activity and mechanism of action of the silver ion in *Staphylococcus aureus* and *Escherichia coli*. Applied and Environmental Microbiology. 2008;74(7):2171–2178. DOI: 10.1128/AEM.02001-07.

[55] Wang Y, Yang Q, Shan G, Wang C, Du J, Wang S, Li Y, Chen X, Jing X, Wei Y. Preparation of silver nanoparticles dispersed in polyacrylonitrile nanofiber film spun by electrospinning. Materials Letters. 2005;59:3046–3049. DOI: 10.1016/j.matlet.2005.05.016

[56] Ucar N, Demirsoy N, Onen A, Karacan I, Kizildag N, Eren O, Vurur OF, Sezer E, Ustamehmetoglu B. The effect of reduction methods and stabilizer (PVP) on the properties of polyacrylonitrile (PAN) composite nanofibers in the presence of nanosilver. Journal of Material Science. 2015;50:1855–1864. DOI: 10.1007/s10853-014-8748-4.

[57] Sichani GN, Morshed M, Amirnasr M, Abedi D. In situ preparation, electrospinning, and characterization of polyacrylonitrile nanofibers containing silver nanoparticles. Journal of Applied Polymer Science. 2010;116:1021–1029. DOI: 10.1002/app.31436.

[58] Kumar R, Munstedt H. Silver ion release from antimicrobial polyamide/silver composites. Biomaterials. 2005;26(14):2081–2088. DOI: 10.1016/j.biomaterials.2004.05.030.

[59] Xu X, Yang Q, Wang Y, Yu H, Chen X, Jing X. Biodegradable electrospun poly(L-lactide) fibers containing antibacterial silver nanoparticles. European Polymer Journal. 2006;42:2081–2087. DOI: 10.1016/j.eurpolymj.2006.03.032.

[60] Jeong L, Park WH. Preparation and characterization of gelatin nanofibers containing silver nanoparticles. International Journal of Molecular Sciences. 2014;15:6857–6879. DOI: 10.3390/ijms15046857.

[61] Zhao XG, Shi JL, Hu B, Zhang LX, Hua ZL. In situ formation of silver nanoparticles inside pore channels of ordered mesoporous silica. Materials Letters. 2004;58(16):2152–2156. DOI: 10.1016/j.matlet.2004.01.022.

[62] Demirsoy N, Ucar N, Onen A, Karacan I, Kizildag N, Eren O, Borazan I. The effect of dispersion technique, silver particle loading, and reduction method on the properties

of polyacrylonitrile–silver composite nanofiber. Journal of Industrial Textiles. 2016;45(6):1173–1187. DOI: 10.1177/1528083714553690.

[63] Rujitanaroj P, Pimpha N, Supaphol P. Preparation, characterization, and antibacterial properties of electrospun polyacrylonitrile fibrous membranes containing silver nanoparticles. Journal of Applied Polymer Science. 2010;116:1967–1976. DOI: 10.1002/app.31498.

[64] Lala NL, Ramaseshan R, Bojun L, Sundarrajan S. Fabrication of nanofibers with antimicrobial functionality used as filters: Protection against bacterial contaminants. Biotechnology and Bioengineering. 2007;97:1357–1365. DOI: 10.1002/bit.21351.

[65] Lee SJ, Heo DN, Moon JH, Ko WK, Lee JB, Bae MS, Park SW, Kim JE, Lee DH, Kim EC, Lee CH, Kwon IK. Electrospun chitosan nanofibers with controlled levels of silver nanoparticles. Preparation, characterization and antibacterial activity. Carbohydrate Polymers. 2014;111:530–537. DOI: 10.1016/j.carbpol.2014.04.026.

[66] Hong KH. Preparation and properties of electrospun poly (vinyl alcohol)/silver fiber web as wound dressings. Polymer Engineering and Science. 2077;47(1):43–49. DOI: 10.1002/pen.20660.

[67] Lee HK, Jeong EH, Baek CK, Youk JH. One-step preparation of ultrafine poly(acrylonitrile) fibers containing silver nanoparticles. Materials Letters. 2005;59:2977–2980. DOI: 10.1016/j.matlet.2005.05.005.

[68] Wang Y, Yang Q, Shan G, Wang C, Du J, Wang S, Li Y, Chen X, Jing X, Wei Y. Preparation of silver nanoparticles dispersed in polyacrylonitrile nanofiber film spun by electrospinning. Materials Letters. 2005;59:3046–3049. DOI: 10.1016/j.matlet.2005.05.016.

[69] Lee DY, Lee KH, Kim BY, Cho NI. Silver nanoparticles dispersed in electrospun polyacrylonitrile nanofibers via chemical reduction. Journal of Sol-Gel Science and Technology. 2010;54:63–68. DOI: 10.1007/s10971-010-2158-0.

[70] Kizildag N, Ucar N. Investigation of the properties of PAN/f-MWCNTs/AgNPs composite nanofibers. Journal of Industrial Textiles. DOI: 10.1177/1528083716632807.

[71] Zhang W, Li W, Wang J, Qin C, Dai L. Composites of polyvinyl alcohol and carbon nanotubes decorated with silver nanoparticles. Fibers and Polymers. 2010;11(8):1132–1136. DOI: 10.1007/s12221-010-1132-3.

[72] Fortunati E, D'Angelo F, Martino S, Orlacchio A, Kenny JM, Armentano I. Carbon nanotubes and silver nanoparticles for multifunctional conductive biopolymer composites. Carbon. 2011;49(7):2370–2379. DOI: 10.1016/j.carbon.2011.02.004.

[73] Grinou A, Bak H, Yun YS, Jin HJ. Polyaniline/silver nanoparticle-doped multiwalled carbon nanotube composites. Journal of Dispersion Science and Technology. 2012;33(5): 750–755. DOI: 10.1080/01932691.2011.567862.

[74] Mi H, Xu Y. Ag–loaded polypyrrole/carbon nanotube: One-step in situ polymerization and improved capacitance. Advanced Materials Research. 2012;531:35–38. DOI: 10.4028/www.scientific.net/AMR.531.35.

[75] Ucar N, Eren O, Onen A, Kizildag N, Demirsoy N, Karacan I. The effect of polyaniline and amine functionalized carbon nanotubes on the properties of composite nanofiber web. Tekstil ve Konfeksiyon. 2014;24(3):266–271.

[76] Eren O, Ucar N, Onen A, Kizildag N, Karacan I. Synergistic effect of polyaniline, nanosilver and carbon nanotube mixtures on the structure and properties of polyacrylonitrile composite nanofiber. Journal of Composite Materials. 2016;50(15):2073–2086. DOI: 10.1177/0021998315601891.

Section 2

Advancement in Electrospinning

Chapter 3

Fabrication of Highly Aligned Poly(Vinyl Alcohol) Nanofibers and its Yarn by Electrospinning

Jeong Hyun Yeum, Seong Baek Yang and
Yeasmin Sabina

Additional information is available at the end of the chapter

http://dx.doi.org/10.5772/65940

Abstract

In the conventional electrospinning method, fibers are randomly deposited and form nonwoven structures; however, highly aligned micro- or nanofiber and its yarn may only be applicable for the fields, including composites, clothing, textiles, and microelectronics. The elementary principle to obtain uniaxially arranged nanofiber array is to regulate the electric field distribution by using assistant electrode or modified collecting device. The potential applications of conventional electrospun poly(vinyl alcohol) (PVA) fiber in the preparation of ultrafine separation filters, biodegradable mats, etc., have been described by many researchers. Highly aligned PVA nanofibers were prepared using a modified electrospinning process at the optimum conditions, and a twister added modified electrospinning apparatus was used to prepare twisted nanofiber yarn. The diameter and arrangement of the electrospun PVA nanofibers were characterized using FE-SEM. To study the effect of applied voltage and rotational velocity on the alignment rate of the nanofibers, different voltages and rotational velocity were applied during modified electrospinning, keeping other parameters unchanged. To measure the melting temperature and crystallinity of aligned nanofibers, differential scanning calorimetry and X-ray diffraction measurement were performed, respectively. The fabricated highly aligned nanofiber and its yarn might have a practical use of devices for microelectronics.

Keywords: poly(vinyl alcohol), aligned nanofiber, yarn, electrospinning, centrifugal jet spinning

1. Introduction

Generally, electrospun nanofibers often exhibit random orientation due to the bending and whipping movement of the electrospinning jet [1]. Control over the unidirectional nanofibers is required for specific applications including composite materials, electrochemical sensing, reinforcements, bone, blood vessel engineering, and tissue engineering [2]. Beside, production of yarn has been of great interest as it is necessary for weaving and knitting. It is more convenient to handle yarn instead of single nanofibers as they are strong enough without the demand of precise and sophisticated equipment [3]. As a result, current interests have been shifted toward the electrospinning of aligned nanofibers and its yarn and some reports are available on them

A facile route was established by Rakesh et al. to control the alignment of electrospun multi-walled carbon nanotube/reinforced poly(vinyl alcohol) (PVA) nanofibers using slotted collector geometries [4]. They showed that the introduction of an insulating into a conductive collector considerably influences the electrostatic forces acting on a charged fiber. Also, they reported in the same study that among different collector geometries, rectangular and slotted collectors with circular ends present good fiber alignment over a large collecting area. Bicomponent-aligned nanofibers of N-carboxyethyl chitosan and PVA were fabricated by Mincheva et al. [5]. Also, they reported in the same paper that 1D-, 1D-transversery, or 3D fiber alignment was fabricated depending on the type of the collector used.

To use the full potential of electrospun nanofibers, it is necessary to assemble into well-ordered structures. Aligned structures nanofibers have also been proven to be better quality for tissue engineering applications, stimulating controlled cell growth, adhesion, and proliferation featuring those of the natural extracellular matrix [6, 7].

Lately, a number of techniques have been developed to fabricate aligned electrospun nanofibrous structures, for example, the use of a rotating mandrel collector, dual-grounded collection plate, a copper wire drum, a scanning tip, a water reservoir collector, dynamic liquid support system, two oppositely metallic spinneret, multiple field, and parallel auxiliary electrodes [8]. The different fiber alignment technique has been used by various research scientists. Zussman et al. reported on a system for producing the ordered fabrication of nanofibers into the crossbar nanostructures. In this technique, the collector disk fitted with a table that can collect the nanofibers and can be revolved about the *z*-axis, each fiber aligned at a set angle to the layer to create layers of nanofiber arrays below, the collector's disk is temporarily stopped and the table revolved the required number of degrees (**Figure 1(A)**) [9]. Katta et al. presented a simple and successful technique for spinning sheets with one centimeter wide strips of aligned nanofibers. Copper wires placed regularly in the form of a round drum as a collector of the electrospun nanofibers was used in this technique and aligned nanofiber layers can be assembled easily without distressing the aligned structure (**Figure 1(B)**) [2]. The conventional method for electrospinning has been modified by Li et al. to fabricate uniaxially aligned nanofibers arrays over larger areas. The reason for the success of this technique was the use of a collector consists of two conductive strips isolated by an insulating gap of flexible width (**Figure 2(A)**) [1]. **Figure 2(B)** and **(C)** show two pairs of conductive bars parted by an insulating material [5]. A facile and reproducible method was reported by Liu et al. in which they established a novel annular collector to arrange the fibrous bundle and identical yarn with well-aligned submicron fibers done by electrospinning, predrafting, and successive twisting (**Figure 3(A)** and **(B)**) [10]. Tamura et al. studied about the fabrication of novel uniaxially aligned ultrafine sulfonated copolyimide nanofibers using an electrospinning apparatus consists of a removable collector made of two conductive aluminum foils and a glass plate insulator (**Figure 4**) [11].

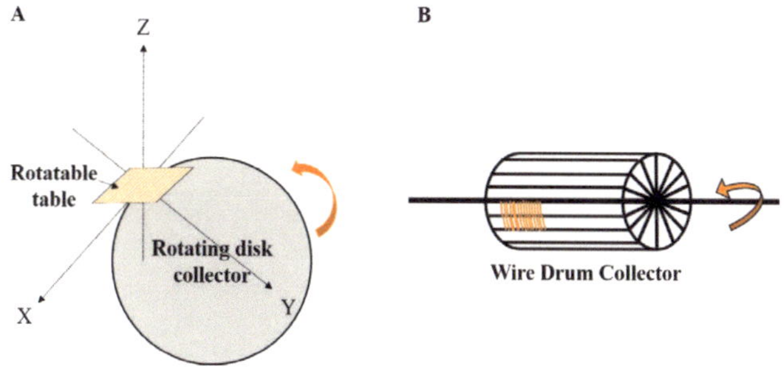

Figure 1. Schematic illustration of the electrospinning method to prepare aligned nanofiber. (A) Showing the collector disk equipped with a table that can collect the nanofiber and can be rotated about the *z*-axis [9] and (B) presenting Plexiglas disk with copper wires that can collect nanofibers as stratified layering [2].

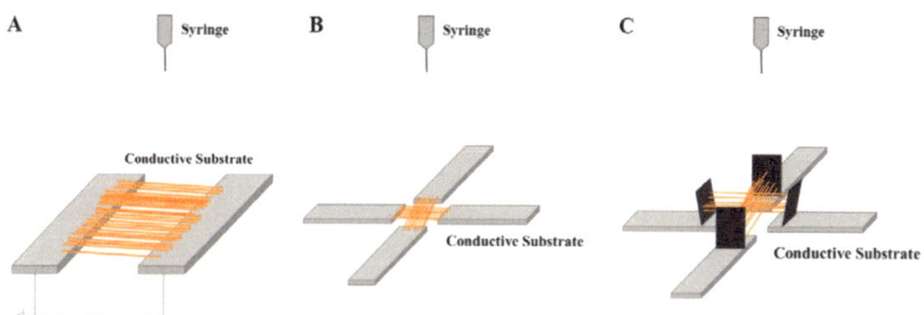

Figure 2. Schematic representation of the collectors used for preparing aligned nanofiber (2A = Ref. [1], 2B and 2C = Ref. [5]).

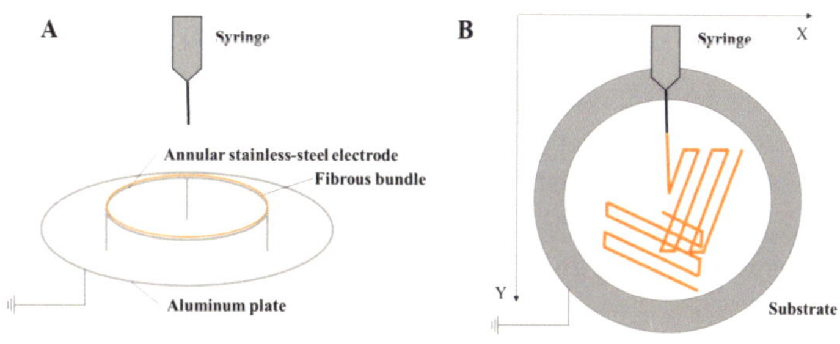

Figure 3. Schematic representation of the electrospinning process for preparing the annular submicron fiber bundle (3A = Ref. [10], 3B = Ref. [27]).

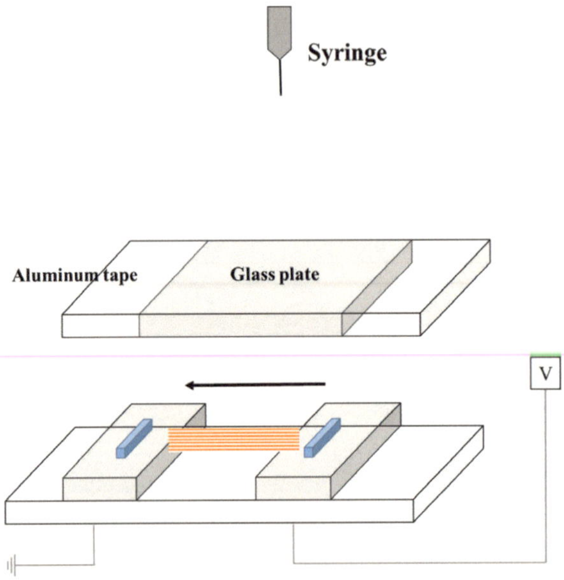

Figure 4. Schematic diagram of the electrospinning method and the collector was composed of two conductive aluminum foils and a glass plate insulator and the collector could be removed from the apparatus [11].

In the beginning, most of the work on electrospinning of nanofiber yarns was actually concentrated on nontwisted nanofiber bundles. Despite twist can be intruded by a postelectrospinning treatment, it is now highly favored that twists can be given directly to continuous nanofiber yarns from an electrospinning process.

In this investigation, we demonstrated a novel parallel collector to obtain the highly aligned PVA nanofibers through modified electrospinning, and a twister added modified electrospinning apparatus was used to prepare twisted nanofiber yarn. The effect of applied voltage and rotational velocity of twister on the alignment rate of the nanofibers were evaluated.

2. Experimental

2.1. Materials

PVA (number-average degree of polymerization of 1700, M_w = 89,000–98,000, fully hydrolyzed, degree of saponification = 99%) was purchased from the DC Chemical Co., Seoul, South Korea. For preparing all solution doubly distilled water was used.

2.2. Preparation of spinning solution

PVA solutions were prepared by dissolving PVA in doubly distilled water under magnetic stirring for 2 h at 80°C followed by cooling to room temperature and various PVA concentrations were used (5 and 7.5 wt.%). For centrifugal jet electrospinning only 5 wt.% PVA was used.

2.3. Electrospinning

For general electrospinning method, PVA solutions were carefully transferred into a syringe and a syringe pump was used to deliver the solution through the blunt needle with a regulated solution feeding rate. Electrospinning was performed under various electric fields (CHUNGPA EMT Co., South Korea), which was applied to the solution using an alligator clip attached to the syringe needle. To collect fibers an electrically grounded aluminum foil was placed at a certain vertical distance from the needle tip. For modified electrospinning method, we used parallel plane collectors to get uniaxially aligned fibers, which consist of two strips or plates of metallic collectors connected to the negative voltage and the needle was connected to the positive voltage. In this method, the electrospinning jet is drawn in between the two plates, ensuring the deposition of uniaxially aligned fibers perpendicular to the plates.

During the electrospinning process, PVA solutions were carefully transferred into a syringe and a syringe pump was used to deliver the solution through the blunt needle with a regulated solution feeding rate. Electrospinning was performed under various electric fields (CHUNGPA EMT Co., South Korea), which was applied to the solution using an alligator clip attached to the syringe needle. To collect fibers an electrically grounded aluminum foil was placed at a certain vertical distance from the needle tip. For getting twisted fibers a rotating twister was attached before the grounded rotating drum collector as shown in **Figure 5**. At first the solidified fibers were twisted by the twister, subsequently accumulated around the rotating drum collector.

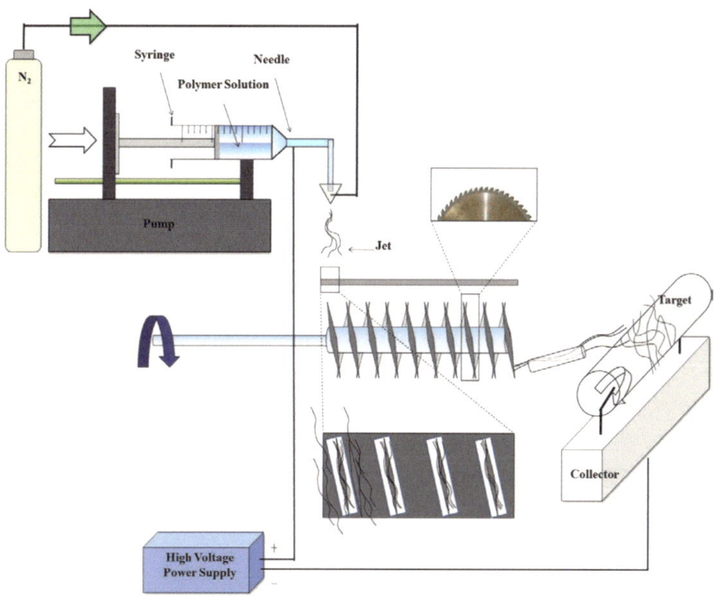

Figure 5. Schematic illustration of a twister added modified electrospinning apparatus to prepare twisted aligned nanofibers yarn.

2.4. Centrifugal jet electrospinning

Generally, the centrifugal jet spinning is composed of motor, disk containing spinneret and collector (**Figure 6**). In this method, aligned nanofibers are fabricated by exploiting high-rotational force, rotating polymer solution jets to eject fiber.

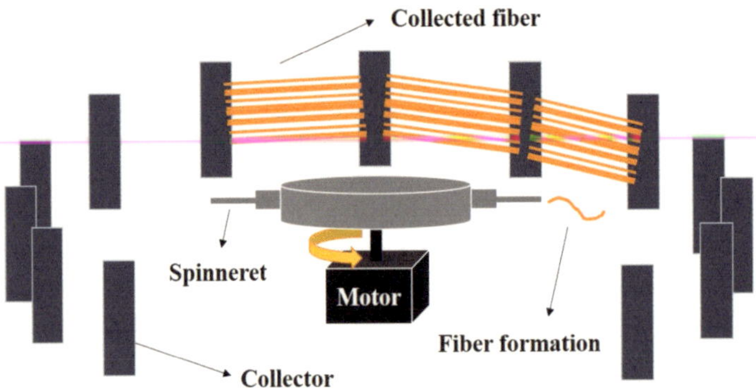

Figure 6. Schematic illustration of the general centrifugal jet spinning apparatus.

In this study, we have made a setup of centrifugal electrospinning on the basis of conventional electrospinning and centrifugal spinning. As shown in **Figure 7**, the syringe is attached to a smaller diameter pipe and fixed vertically above a circular rotating disk containing many metallic needles like ejection path for polymer solution. The disk is attached to a speed-adjustable motor and TCD is 15 cm. A round case-like collector made of metallic wire placed around the rotating disk for collecting the nanofibers. The high voltage was applied between the ejection needle and collector. The polymeric jet originated from the ejection end under the action of electrical force and centrifugal force and the jets are extended by the repulsive force of charges on the jets [12]. Highly aligned ultrafine PVA nanofibers are fabricated after the jets solidify and fall onto the collector.

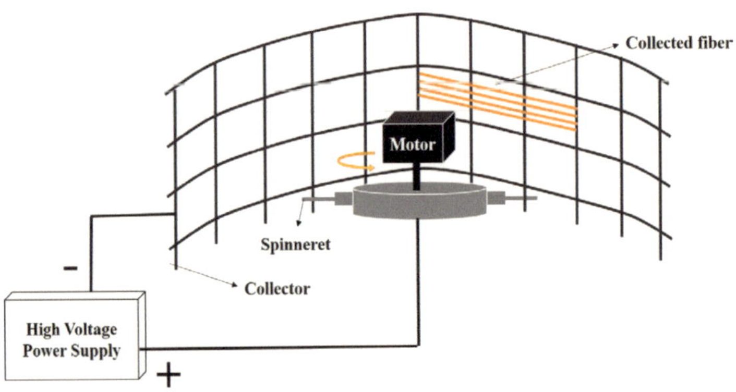

Figure 7. Schematic illustration of the centrifugal jet electrospinning apparatus.

2.5. Characterization

The morphology of PVA nanofibers and its yarn were examined using a FE-SEM (SU8220, Hitachi, Japan). From the FE-SEM images the fiber diameter was measured. Photoshop 7 was used to measure the average fiber diameter and at least 20 different fibers and 100 different segments were randomly selected from each image. The degree of nanofibers alignment was studied against different voltages keeping other parameters constant. To measure the melting temperature and crystallinity of aligned nanofibers differential scanning calorimetry and X-ray diffraction measurement were performed, respectively. To determine the degree of nanofiber alignment, several FE-SEM images were captured from each sample.

3. Results and discussion

3.1. Fabrication of highly aligned PVA nanofibers by electrospinning

PVA is a well-known hydrophilic, semicrystalline polymer and has received great attention due to its good chemical resistance, good thermal stability, good physical properties, excellent biocompatibility, and low price [13, 14]. The potential applications of electrospun PVA fiber in the preparation of ultrafine separation filters, biodegradable mats, etc. have been described by many researchers [14, 15]. As a simple and versatile process electrospinning is offering unique capabilities for preparing fibers from polymer solutions with diameters ranging from the nano- to microscale which generated interesting applications in fields of protective clothing, filtration, drug delivery, self-cleaning, tissue engineering, electronic and photonic devices, etc. [16, 17]. However, the function is somewhat limited so far, as the maximum electrospun fibers are in the form of isotropic nonwoven mats [18]. As a result current interests have been shifted toward electrospinning-aligned nanofiber yarn for the purpose of use in microelectronics, photonics, and in a variety of electrical, optical, mechanical, and biomedical applications [8]. Due to the high surface area, porous structure, and functional molecules and nanomaterials adopting ability (e.g., nanoparticles and nanotubes), nanofibers have been used in various areas including batteries, biomedical, sensors, fuel cells, nanocomposites, and protective clothing [19–25]. Due to the "whipping instability" of the electrospinning jet, typical electrospun nanofibers are usually collected as nonwoven or randomly oriented structures, since disordered orientation and low mechanical strength of the fibrous structure have restricted their applications, early studies were focused on regulating the fiber alignment.

Aligned fibers are considered as potential candidates for the design of directional optical components, for instance waveguides and lasers [26]. To obtain aligned electrospun fibers, several approaches have been established, however, parallel electrode method has gained more attention due to the higher degree of alignment of the fibers [26, 27]. In parallel electrode method, aligned nanofibers are obtained through the manipulation of the electric field that was manipulated by using two parallel plates (gap collector) joined with grounded electrode [1, 28]. It is possible to get a high degree of fiber alignment using this method, and the produced fibers can be easily applicable for device fabrication.

In this section, we are reporting about the preparation of highly aligned PVA nanofibers using a modified electrospinning apparatus composed of two parallel vertical plate collectors. The diameter and arrangement of the electrospun PVA nanofibers were characterized using a field emission scanning electron microscope (FE-SEM). The degree of nanofiber alignment was studied against different voltages keeping other parameters constant. The melting temperature and crystallinity of aligned nanofibers with different voltages were also studied.

3.1.1. Morphological differences of nanofibers

Figure 8(A) and **(b)** show the FE-SEM images of electrospun nanofibers prepared by the general electrospinning method and modified electrospinning method, respectively. It is found that although smooth, defect-free and round-shaped fibers are collected using both the modified and unmodified electrospinning method, the alignment of the nanofibers is very different for these two methods. Manipulation of the electric fields that were manipulated by using two parallel vertical plates connected with positive voltage coupled with the repelling force from the residual charges on the electrospun nanofibers ensure good alignment of the electrospun nanofiber as shown in **Figure 8(B)**. In contrast, a regular oriented distribution of electrospun fibers cannot be achieved using unmodified collector as shown in **Figure 8(A)**.

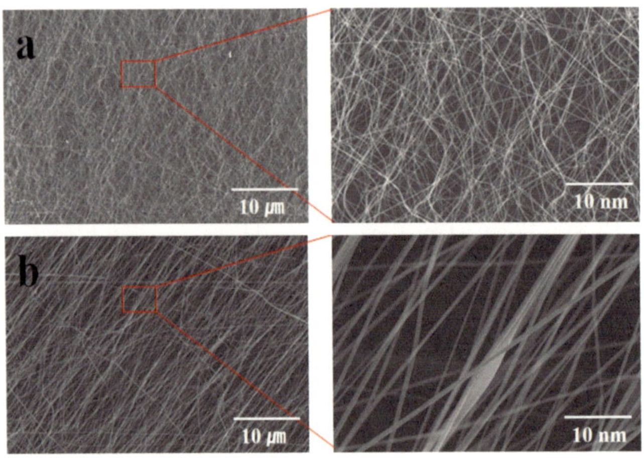

Figure 8. FE-SEM images of PVA nanofiber prepared by (a) general electrospinning method and (b) modified electrospinning method. (Concentration of PVA solution = 7.5 wt.%, TCD = 15 cm, and applied voltage = 10 kV.)

The FE-SEM-measured nanofiber diameter distributions for both preparation methods are shown in **Figure 9** for the purpose of comparison. In the case of the nonaligned PVA nanofibers prepared using general electrospinning methods (**Figure 9(A)**), the average diameter value is $\langle D \rangle$ = 243.7 nm with a standard deviation of σ = 82.82 for N = 100 measurement. In case of aligned nanofiber prepared by modified electrospinning method (**Figure 9(B)**) value of $\langle D \rangle$ = 1013.9 nm and σ = 613.2 were estimated for N = 100 measurement. These statistics illustrated that a better control over nanofiber diameter was gained with the nonaligned PVA nanofibers compared to aligned nanofiber.

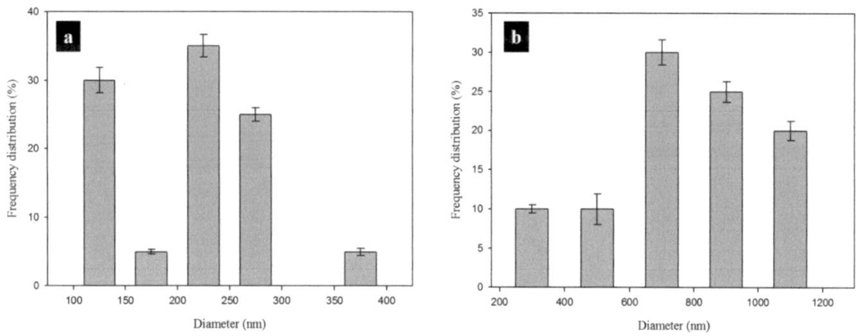

Figure 9. Diameter distribution of PVA nanofiber prepared by (a) general electrospinning method and (b) modified electrospinning method. (Concentration of PVA solution = 7.5 wt.%, TCD = 15 cm, and applied voltage = 10 kV.)

3.1.2. Effect of applied voltage on melting temperature and crystallinity of nanofiber

The melting temperature and percentage of crystallinity of aligned PVA nanofiber prepared by modified electrospinning method at various voltages are presented in **Figure 10**. As can be seen, applied voltage plays a pivotal role on both melting temperature and crystallinity rate of aligned PVA nanofiber. The results showed that both melting temperature and crystallinity rate increase as the applied voltage increased. For example, consider the value of melting temperature and crystallinity rate at voltage 10, 12, 14, 16, 18, and 20 kV. The numerical values for the melting temperature are 223, 223.8, 224.6, 225.2, 226, and 227°C and corresponding values for crystallinity rate are 50, 50.4, 51.5, 52.1, 52.8, and 53.2%, respectively. From the results it is clear that the higher melting temperature and crystallinity rate of aligned PVA nanofiber can be obtained at applied voltage 20 kV.

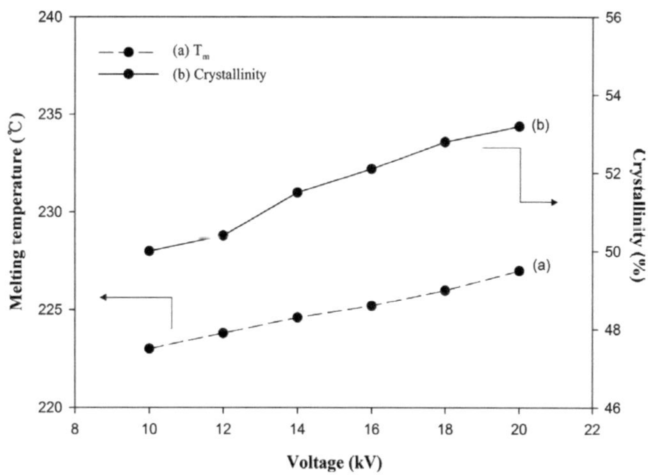

Figure 10. Effect of voltage on melting temperature and degree of crystallinity of aligned PVA nanofiber. (PVA solution concentration = 5 wt.%, TCD = 10 cm).

3.2. Preparation-aligned PVA nanofiber yarn using a twister added modified electrospinning method

One-dimensional nanomaterials having diameters less than 1 μm (1000 nm), and an aspect ratio (length/diameter) larger than 100:1, are defined as nanofibers or superfine or ultrathin fibers. Nanofibers are also called as submicron fibers when they are in the range of 100–1000 nm [29–31]. As we know, a conventional electrospinning process requires the application of electrostatic force between polymer solution kept in a syringe and a counter metal electrode, for example, as a plate or a rotating drum placed at a suitable distance. The electrostatic forces overcome the surface tension of the polymer solution if sufficient high electrical field is applied, as a result a thin jet can be ejected from the drop of polymer solution formed at the tip of the spinneret. At first the charged jet undergoes a stable stretching, however, soon it starts bending and whipping randomly due to further stretching of the jet and evaporation of solvent [32, 33]. It is possible to collect electrospun fibers as oriented parallel to the direction of rotation [34]. Yarns can be interlaced into many fibrous structures by various processes; however, single nanofibers being fragile and thin are not appropriate for such processes, and sophisticated equipment is required for exact handling of these fibers [35].

Various attempts had been taken by the researchers to obtain nanofibers yarn. Formhals attempted a method to produce nanofiber yarns from electrospun nanofibers and reported in 1934. He described the experimental setups of producing polymer filaments using electrostatic forces in a series of patents. In the beginning, most of the works on electrospinning of nanofiber yarns were actually concentrated on nontwisted nanofiber bundles. Despite twist can be intruded by a postelectrospinning treatment, it is now highly favored that twists can be given directly to continuous nanofiber yarns from an electrospinning process. The deposition of nanofibers in the form of a narrow strip using a series of charged rings in the electrospinning zone was demonstrated by Deitzel et al. in 2001, both the charged rings and the surface charge on the jet had the same polarity that can enhance the downward force on the jet resulted in nanofiber deposition in a narrow strip (0.6 cm wide) on the target of rotating drum [36]. Theron et al. [37] demonstrated an electrostatic field-assisted assembly technique in combination with the dynamic rotating collector to create individual nanofibers positioned and aligned. In this method, a tapered wheel-like disk made of aluminum was used to collect bundles of nanofibers. The jet emerged from the droplet, at first form an envelope cone after that shrink to form an inverted cone, once it reached the sharp-tapered edge of the wheel collector which had a strong converging electric field. Accordingly, nanofibers would preferentially deposit on the wheel edges, which were accumulated in a parallel array due to the rotation of the wheel. Teo and Ramakrishna [38] described the fabrication of nanofiber bundles by collecting them across two negatively charged steel blades followed by dipping them in water that was thus required to assemble the fibers together by the force of water surface tension. The resulting fiber bundles can be transferred into other substrates, and to be twisted or braided manually.

In this section, we are reporting about a novel twister added self-bundling electrospinning method for generating continuous twisted aligned PVA electrospun fiber yarn. Compared with conventional electrospinning setup, the special thing in this process is that a twister is used to induce the self-bundling of polymer nanofibers before deposited on the collector. The diameter

http://dx.doi.org/10.5772/65940

and arrangement of the electrospun PVA nanofibers were characterized using FE-SEM. The degree of nanofibers alignment was studied against different voltages and rotational velocity of twister keeping other parameters constant.

3.2.1. PVA-aligned nanofiber yarn morphology

Figure 11 represents the FE-SEM image of the twisted yarn of PVA prepared using rotational velocity 25 rpm, applied voltage 15 kV, TCD 15 cm, and PVA solution concentration at 5 wt.%. For the following study, we have prepared the twisted yarn at low rotational velocity. As can be seen most of the filaments included ae twisted. As showing in **Figure 11**, the fiber surface reveals that highly twisted, smooth, uniform with a higher alignment rate were obtained.

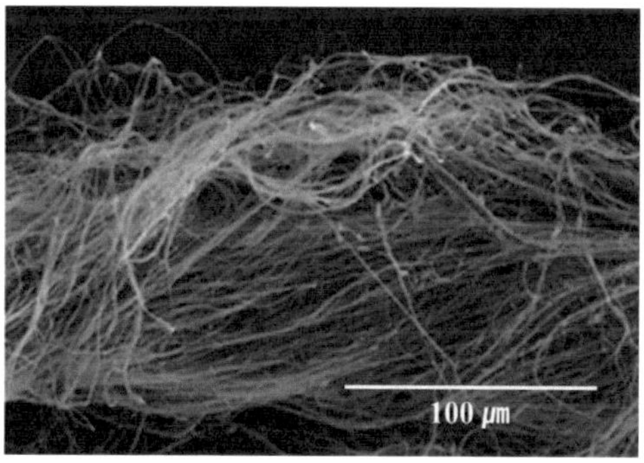

Figure 11. FE-SEM image of twisted aligned PVA nanofiber yarn prepared by twister added modified electrospinning apparatus. (PVA solution concentration= 5 wt.%, TCD = 15 cm, applied voltage = 15 kV, rotational velocity = 25 rpm).

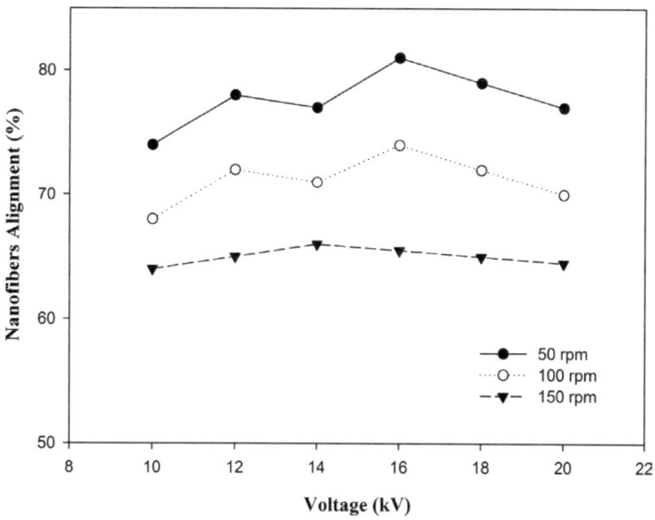

Figure 12. Effect of voltage and rotational velocity on the degree of alignment of twisted PVA nanofiber.

3.2.2. Effects of voltages and rotational velocity on degree of nanofiber alignment in yarn

During electrospinning process, voltage and rotational velocity of twister play significant role in nanofiber alignment rate. A series of experiments were carried out by changing the applied voltage and rotational velocity in the range of 10–20 kV and 50–150 rpm, respectively (**Figure 12**). As illustrated, voltage has no significant effect on fiber alignment rate, however, considerable change was found on nanofiber alignment at various rotational velocities. We measured nanofiber alignment rate against three different rotational velocities of twister including 50, 100, and 150 rpm at various voltages in the range of 8–20 kV. A higher degree of nanofiber alignment was found at the rotational velocity of 150 rpm. In contrast, the lowest degree of nanofiber alignment was obtained at 50 rpm. It can be also seen that the maximum degree of nanofiber alignment was obtained at 150 rpm and 16 kV.

3.2.3. Effects of rotational velocity on nanofiber morphology in yarn

Figure 13 shows several FE-SEM images of various samples taken from the twister-added rotating drum collector. Several sets of experiments were conducted at various rotational velocities of twister, and images were captured to confirm twisting and alignment. To acquire different perspective, images were taken at various magnifications. As seen in the FE-SEM images, well-aligned, highly twisted and thin nanofibers are present at 150 rpm. And the possible reason of smaller diameter nanofiber is the removal of residual solvent from the nanofibers due to higher rotational velocity. Beside, beads are seen at 100 and 150 rpm. Finally, it can be concluded that uniform, highly twisted, and well-aligned nanofiber with less beads defects are only achievable at 150 rpm.

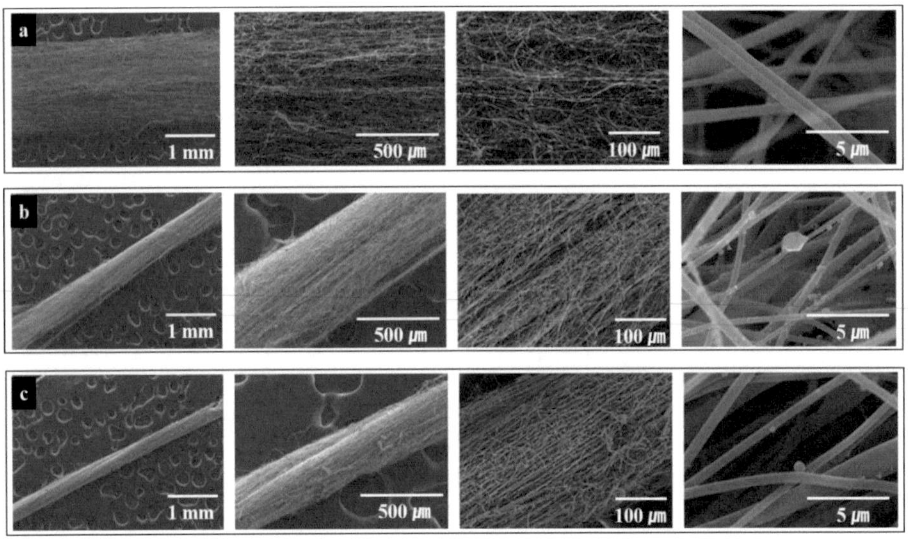

Figure 13. FE-SEM images of aligned twisted PVA nanofiber yarn prepared at different rotation velocity (a) 50 rpm, (b) 100 rpm, and (c) 150 rpm. (PVA solution concentration = 7.5 wt.%, TCD = 15 cm, and applied voltage = 16 kV).

3.2.4. Effects of rotational velocity on diameter distribution of nanofiber in yarn

The diameter distributions of aligned PVA nanofiber yarn gained at three different rotational velocities of twister were given in **Figure 14**. We considered three rotational velocities of twister including 50, 100, and 150 rpm. Smaller average fiber diameter and a large amount of thin fibers with diameter bellow 400 nm were found at 150 rpm (**Figure 14(C)**). On the other hand, a narrower distribution of fiber diameters was observed at 100 rpm (**Figure 14(B)**), however, beaded nanofiber with lower alignment rate was observed at this rotational velocity (**Figure 13**). Comparatively less control in the fiber diameter with average diameter 444 nm was seen at 50 rpm.

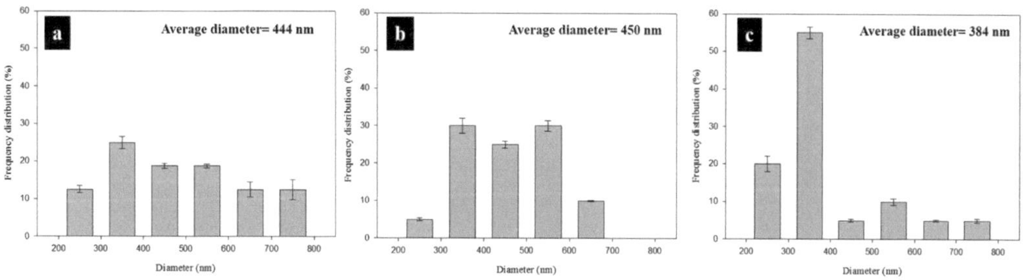

Figure 14. Diameter distribution of aligned twisted PVA nanofiber yarn prepared at different rotation velocity (a) 50 rpm, (b) 100 rpm, and (c) 150 rpm. (PVA solution concentration = 7.5 wt.%, TCD = 15 cm, and applied voltage = 16 kV).

3.2.5. Effects of rotational velocity on nanofiber morphology in yarn

The FE-SEM images of the PVA yarn collected at 200 rpm are presented in **Figure 15**. Rotation velocity of twister has considerable effect on nanofiber diameter [39]. From the magnified FE-SEM image (**Figure 15(C)**), it is clear that smaller diameter and well-aligned nanofibers were fabricated. In addition, it is also seen that smooth, defects free, and round shape nanofibers exist.

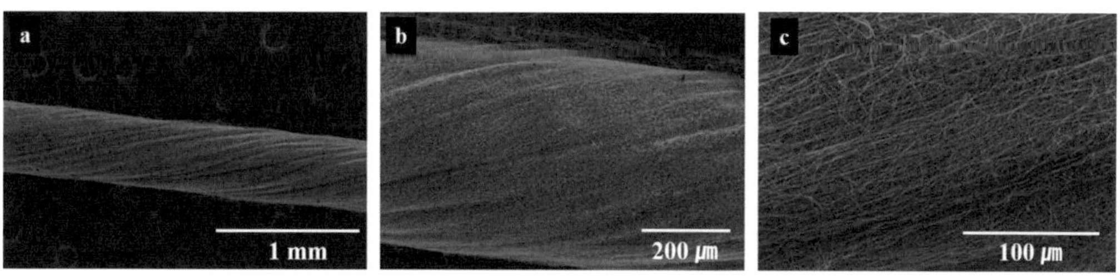

Figure 15. FE-SEM images of (a) aligned twisted PVA nanofiber yarn prepared at rotational velocity of 200 rpm. The images (b) and (c) are magnified form of image (a). (PVA solution concentration = 7.5 wt.%, TCD = 15 cm, and applied voltage = 16 kV).

3.3. Fabrication of highly aligned PVA nanofibers by centrifugal jet electrospinning

Recently, centrifugal jet spinning technique has drawn much attention of the scientists for the production of nanosized fiber due to its amazing characteristics, for example, large-scale production rate, facile, highly efficient, low cost fabrication technique, etc. [40]. The method uses centrifugal forces together with the viscoelastic properties and the mass transfer characteristics of spinning solutions to support the controlled thinning of a polymer solution filament into nanofibers [41].

Many researchers have tried centrifugal jet spinning method to prepare aligned nanofiber for various purposes. Badrossamay et al. [42] studied a facile technique for fabricating aligned three-dimensional nanofiber structures by exploiting high-speed, rotating polymer solution jets to eject fiber. They also demonstrated the usefulness of this technique for building uniaxially aligned nanofiber structures necessary for tissue engineering. A novel and effective centrifugal electrospinning setup with rotating polymer solution jets was demonstrated by Liu et al. to prepare uniaxially aligned and cross-aligned arrays of ultrafine polymer fibers [43]. Amalorpava Mary et al. had attempted a novel method to obtain centrifugal spun fibrous web as a drug [44]. On considering the huge prospective and effectiveness of this technique, we have also made some attempts to make centrifugal spun nanofibers.

The centrifugal jet spinning system was composed of a DC hobby motor (9–18 V, Radio Shack) powered hollow chamber with two orifices passing through the chamber wall and a set of product collecting posts. Teflon tubes (outer diameter of 15 mm and an inner diameter of 10 mm) were used to make spinning chamber. The diameter of the orifices present in the wall of the spinning chamber was 500 μm. It is possible to adjust the distance between the center of the chamber and the product collecting posts. At a particular flow rate, spinning solutions were continuously fed into the chamber and the centrifugal force exceeded the capillary force of the spinning solution in the orifice as the spinning chamber spins. Consequently, spinning solution jet is ejected from the rotating chamber and by adjusting the voltage applied on the motor the rotational speed of the chamber can be changed [41].

In this section, we are reporting about the fabrication of highly aligned PVA nanofibers by a unique and efficient centrifugal electrospinning setup with rotating polymer solution jets. The developed fibers were morphologically characterized by FE-SEM and the results showed that highly aligned PVA nanofibers were successfully assembled.

3.3.1. Morphology of centrifugal jet electrospun-aligned nanofibers

Figure 16 shows the FE-SEM pictures of highly aligned PVA nanofibers (PVA solution concentration was 5 wt.%), which were obtained at the optimized conditions (TCD = 15 cm, applied voltage = 15 kV, and rotational velocity = 5000 rpm). To get clear idea images were taken at various magnifications. As seen in the FE-SEM images excellent alignment is present. From **Figure 16(D)**, it is clear that well-aligned nanofibers with higher alignment rate can be produced through our new centrifugal jet electrospinning setup. Excellent properties of nanofibers are observed, including uniform, feed free, and smooth surface.

Figure 16. FE-SEM pictures of aligned nanofiber prepared by centrifugal jet electrospinning method. (Images b, c, and d are magnified form of image a.) (PVA solution concentration = 5 wt.%, TCD = 15 cm, applied voltage = 15 kV, rotational velocity = and 5000 rpm).

4. Conclusion

The study was aimed to apply a modified collector composed of two parallel vertical plates, and a twister-added collector to prepare highly aligned PVA nanofiber and its yarn, respectively. The result showed that this process was facile to prepare highly aligned PVA nanofiber and its yarn. In addition, a good reproducibility could be gained. The diameter and arrangement of the electrospun PVA-aligned nanofibers and its yarn were characterized using FE-SEM. For PVA-aligned nanofiber, we also studied the effect of voltage on alignment rate, melting temperature, and crystallinity. The results showed melting temperature and crystallinity increase gradually as the applied voltage increased, and the maximum degree of alignment was achieved at applied voltage 20 kV. On the other hand, for PVA yarn, effect of applied voltage and rotational velocity of twister on nanofiber alignment rate were examined and found that applied voltage 16 kV and rotational velocity 150 rpm are favorable conditions for getting highly aligned yarn. In addition, the effect of rotational velocity on diameter distribution and morphology of PVA nanofiber were also tested. The FE-SEM photographs revealed that smooth, well aligned, and bead free with good control on fiber diameter were obtained at the rotational velocity of 150 rpm. Beside aligned ultrafine PVA nanofibers have also been successfully assembled. The results indicated that centrifugal electrospinning is a competent technique to assemble highly aligned ultrafine PVA nanofibers which have a great prospective to use in electronic and optoelectronic devices.

Acknowledgements

This study was supported by Basic Science Research Program through the National Research Foundation of Korea (NRF) funded by the Ministry of Education (2011–0009602). In addition, this study was partially supported by the National Research Foundation of Korea (NRF) funded by the Ministry of Science, ICT, and Future Planning (2016R1A2B4010329).

Author details

Jeong Hyun Yeum*, Seong Baek Yang and Yeasmin Sabina

*Address all correspondence to: jhyeum@knu.ac.kr

Department of Bio-fibers and Materials Science, Kyungpook National University, Daegu, Republic of Korea

References

[1] Li D, Wang Y, Xia Y. Electrospinning nanofibers as uniaxially aligned arrays and layer-by-layer stacked films. Advanced Materials. 2004;16(4):361–366.

[2] Katta P, Alessandro M, Ramsier RD, Chase GG. Continuous electrospinning of aligned polymer nanofibers onto a wire drum collector. Nano Letters. 2004;4(11):2215–2218. DOI: 10.1021/nl0486158

[3] Yousefzadeh M, Latifi M, Teo WE, Amani-Tehran M, Ramakrishna S. Producing continuous twisted yarn from well-aligned nanofibers by water vortex. Polymer Engineering & Science. 2011;51(2):323–329. DOI: 10.1002/pen.21800

[4] Rakesh GR, Ranjit GS, Karthikeyan KK, Radhakrishnan P, Biji P. A facile route for controlled alignment of carbon nanotube-reinforced, electrospun nanofibers using slotted collector plates. eXPRESS Polymer Letters. 2015;9(2):105–118. DOI: 10.3144/expresspolymlett.2015.12.

[5] Mincheva R, Manolova N, Rashkov I. Bicomponent aligned nanofibers of N-carboxyethylchitosan and poly(vinyl alcohol). European Polymer Journal. 2007;43(7):2809–2818. DOI: 10.1016/j.eurpolymj.2007.04.020

[6] Xu CY, Inai R, Kotaki M, Ramakrishna S. Aligned biodegradable nanofibrous structure: a potential scaffold for blood vessel engineering. Biomaterials. 2004;25(5):877–886. DOI: 10.1016/S0142-9612(03)00593-3

[7] Lee CH, Shin HJ, Cho IH, Kang YM, Kim IA, Park KD, Shin JW. Nanofiber alignment and direction of mechanical strain affect the ECM production of human

ACL fibroblast. Biomaterials. 2005;26(11):1261–1270. DOI: 10.1016/j.biomaterials.2004.04.037

[8] Wang X, Zhang K, Zhu M, Yu H, Zhou Z, Chen Y, Hsiao BS . Continuous polymer nanofiber yarns prepared by self-bundling electrospinning method. Polymer. 2008;49(11):2755–2761. DOI: 10.1016/j.polymer.2008.04.015

[9] Zussman E, Theron A, Yarin AL. Formation of nanofiber crossbars in electrospinning. Applied Physics Letters. 2003;82(6):973–975.

[10] Liu CK, Sun RJ, Lai K, Sun CQ, Wang YW. Preparation of short submicron-fiber yarn by an annular collector through electrospinning. Materials Letters. 2008;62(29):4467–4469. DOI: 10.1016/j.matlet.2008.07.058

[11] Tamura T, Takemori R, Kawakami H. Proton conductive properties of composite membranes containing uniaxially aligned ultrafine electrospun polyimide nanofiber. Journal of Power Sources. 2012;217:135–141. DOI: 10.1016/j.jpowsour.2012.05.118

[12] Reneker DH, Yarin AL. Electrospinning jets and polymer nanofibers. Polymer. 2008;49(10):2387–2425. DOI: 10.1016/j.polymer.2008.02.002

[13] Pitt S, Surawut C. On the electrospinning of poly(vinyl alcohol) nanofiber mats: a revisit. Journal of Applied Polymer Science. 2008;108:969–978. DOI: 10.1002/app.27664

[14] Koshi A, Yim K, Shivkumar S. Effect of molecular weight on fibrous PVA produced by electrospinning. Materials Letters. 2003;58(3–4):493–497. DOI: 10.1016/S0167-577X(03)00532-9

[15] Yao L, Haas T, Guiseppi-Elie A, Gary L, Simpson D, Wnek G. Electrospinning and stabilization of fully hydrolyzed poly(vinyl alcohol) fibers. Chemistry of Materials. 2003;15(9):1860–1864. DOI: 10.1021/cm0210795

[16] Ramakrishna S, Fujihara K, Teo W, Yong T, Ma Z, Ramaseshan R. Electrospun nanofibers: solving global issues. In: S. Bland, editors. Materials Today. 9th ed. Oxford, United Kingdom: Elsevier Ltd.; 2006. pp. 40–50. DOI: 10.1016/S1369-7021(06)71389-X

[17] Qin X, Wang S. Filtration properties of electrospinning nanofibers. Journal of Applied Polymer Science. 2006;102(2):1285–1290. DOI: 10.1002/app.24361

[18] Shin YM, Hohman MM, Brenner MP, Rutledge GC. Experimental characterization of electrospinning: the electrically forced jet and instabilities. Polymer. 2001;42(25):09955–09967. DOI: 10.1016/S0032-3861(01)00540-7

[19] Norris ID, Shaker MM, Ko FK, Macdiarmid AG. Electrostatic fabrication of ultrafine conducting fibers: polyaniline/polyethylene oxide blends. Synthetic Metals. 2000;114(2):109–114. DOI: 10.1016/S0379-6779(00)00217-4

[20] Liang D, Hsiao BS, Chu B. Functional electrospun nanofibrous scaffolds for biomedical applications. Advanced Drug Delivery Reviews. 2007;59(14):1392–1412. DOI: 10.1016/j.addr.2007.04.021

[21] Agarwal S, Wendorff JH, Greiner A. Use of electrospinning technique for biomedical applications. Polymer. 2008;49(26):5603–5621. DOI: 10.1016/j.polymer.2008.09.014

[22] Wang X, Drew C, Lee SH, Senecal KJ, Kumar J, Samuelson LA. Electrospinning technology: a novel approach to sensor application. Journal of Macromolecular Science, Pure and Applied Chemistry. 2002;A39(10):1251–1258. DOI: 10.1081/MA-120014850

[23] Shabani I, Hasani-Sadrabadi MM, Haddadi-Asl V, Soleimani M. Nanofiber-based polyelectrolytes as novel membranes for fuel cell applications. Journal of Membrane Science. 2011;368(1–2):233–240. DOI: 10.1016/j.memsci.2010.11.048

[24] Huang ZM, Zhang YZ, Kotaki M, Ramakrishna S. A review on polymer nanofibers by electrospinning and their applications in nanocomposites. Composites Science and Technology. 2003;63(15):2223–2253. DOI: 10.1016/S0266-3538(03)00178-7

[25] Lee S, Obendorf SK. Use of electrospun nanofiber web for protective textile materials as barriers to liquid penetration. Textile Research Journal. 2007;77(9):696–702.

[26] Ner Y, Asemota C, Olson JR, Sotzing GA. Nanofiber alignment on a flexible substrate: hierarchical order from macro to nano. ACS Applied Materials & Interfaces. 2009;1(10): 2093–2097. DOI: 10.1021/am900382f

[27] Fuh YK, Chen Z, He ZY. Direct-write, highly aligned chitosan-poly(ethylene oxide) nanofiber patterns for cell morphology and spreading control. Nanoscale Research Letters. 2013:8(1):97/1–97/9. DOI: 10.1186/1556-276X-8-97

[28] Kameoka J, Czaplewski D, Liu H, Craighead HG. Polymeric nanowire architecture. Journal of Materials Chemistry. 2004;14(10):1503–1505.

[29] Bellan LM, Coates GW, Craighead HG. Poly(dicyclopentadiene) submicron fibers produced by electrospinning. Macromolecular Rapid Communications. 2006;27(7): 511–515.

[30] Kwon IK, Kidoaki S, Matsuda T. Electrospun nano- to microfiber fabrics made of biodegradable copolyesters: structural characteristics, mechanical properties and cell adhesion potential. Biomaterials. 2005;26(18):3929–3939. DOI: 10.1016/j.biomaterials. 2004.10.007

[31] Givens SR, Gardner KH, Rabolt JF, Chase DB. High-temperature electrospinning of polyethylene microfibers from solution. Macromolecules. 2007;40(3):608–610. DOI: 10.1021/ma062398a

[32] Thompson CJ, Chase GG, Yarin AL, Reneker DH. Effects of parameters on nanofiber diameter determined from electrospinning model. Polymer. 2007;48(23):6913–6922. DOI: 10.1016/j.polymer.2007.09.017

[33] Tripatanasuwan S, Zhong Z, Reneker DH. Effect of evaporation and solidification of the charged jet in electrospinning of poly(ethylene oxide) aqueous solution. Polymer. 2007;48(19):5742–5746. DOI: 10.1016/j.polymer.2007.07.045

[34] Doshi J, Reneker DH. Electrospinning process and applications of electrospun fibers. Journal of Electrostatics. 1995;35(2&3):151–160. DOI: 10.1016/0304-3886(95)00041-8

[35] Tan EPS, Ng SY, Lim C. Tensile testing of a single ultrafine polymeric fiber. Biomaterials. 2005;26(13):1453–1456. DOI: 10.1016/j.biomaterials.2004.05.021

[36] Deitzel JM, Kleinmeyer JD, Hirvonen JK, Beck Tan NC. Controlled deposition of electrospun poly(ethylene oxide) fibers. Polymer. 2001;42(19):8163–8170. DOI: 10.1016/S0032-3861(01)00336-6

[37] Theron A, Zussman E, Yarin AL. Electrostatic field-assisted alignment of electrospun nanofibres. Nanotechnology. 2001;12(3):384–390.

[38] Teo WE, Ramakrishna S. Electrospun fibre bundle made of aligned nanofibres over two fixed points. Nanotechnology. 2005;16(9):1878–1884. DOI: 10.1088/0957-4484/16/9/077

[39] Jose MV, Steinert BW, Thomas V, Dean DR, Abdalla MA, Price G, Janowski GM. Morphology and mechanical properties of Nylon 6/MWNT nanofibers. Polymer. 2007;48(4):1096–1104. DOI: 10.1016/j.polymer.2006.12.023

[40] Ren L, Kotha SP. Centrifugal jet spinning for highly efficient and large-scale fabrication of barium titanate nanofibers. Materials Letters. 2014;117:153–157. DOI: 10.1016/j.matlet.2013.11.103

[41] Ren L, Ozisik R, Kotha SP, Underhill PT. Highly efficient fabrication of polymer nanofiber assembly by centrifugal jet spinning: process and characterization. Macromolecules. 2015;48(8):2593–2602. DOI: 10.1021/acs.macromol.5b00292

[42] Badrossamay MR, McIlwee HA, Goss JA, Parker KK. Nanofiber assembly by rotary jet-spinning. Nano Letters. 2010;10(6):2257–2261. DOI: 10.1021/nl101355x

[43] Liu SL, Long YZ, Zhang ZH, Zhang HD, Sun B, Zhang JC, Han WP. Assembly of oriented ultrafine polymer fibers by centrifugal electrospinning. Journal of Nanomaterials. 2013;2013:713275–713284. DOI: 10.1155/2013/713275

[44] Amalorpava Mary L, Senthilram T, Suganya S, Nagarajan L, Venugopal J, Ramakrishna S, Giri DVR. Centrifugal spun ultrafine fibrous web as a potential drug delivery vehicle. eXPRESS Polymer Letters. 2013;7(3):238–248. DOI: 10.3144/expresspolymlett.2013.22

Chapter 4

Dependent and Independent Parameters of Needleless Electrospinning

Fatma Yalcinkaya , Baturalp Yalcinkaya and
Oldrich Jirsak

Additional information is available at the end of the chapter

http://dx.doi.org/10.5772/65838

Abstract

Electrospinning is a simple method to produce nanofibers from solutions and melt of different polymers and polymer blends. There is an extensive application in future of Electrospun nanofibers. Several methods for the production of nanofibers have been developed for their wide-scale production. In this chapter, we introduced the needleless roller electrospinning system that is well known under the trade name of nanospider and used as industrial production scale during the last decade. For industrial production, it is crucial to control and the measure all the spinning parameters of the needleless electrospinning system. Herein, all the electrospinning parameters of the needleless roller electrospinning system were determined and grouped as dependent and independent parameters. Each parameter was defined, and some experimental results are shown under their group. Using theoretical calculations, the minimum electrical field to start initiation of Taylor's cone and the dimensionless electrospinning number was demonstrated. The dimensionless electrospinning number is important for the initiation of the electrospinning system. Each parameter explained in detail, and measurement methods of parameters were clarified. It was found that each parameter plays a major role in productivity and quality of nanofiber webs. Changing the dependent and independent parameters of the electrospinning, the fiber morphology can be adjusted according to demands.

Keywords: roller electrospinning, parameters, nanofiber, needleless electrospinning

1. Introduction

Electrospinning is a versatile method to produce ultra-thin fibers less than 1 μm. Nanofibers produced from electrospinning have enormous properties such as high surface area, highly pore structure, small pore size, and electrical properties, and so on. Based on specific properties of nanofibers, scientists focus on a promising application area of nanofibers. Nanofibers can be used for filtration, wound dressing, drug delivery, tissue engineering, artificial vessels, barrier textiles, sensors, carriers, and so on [1–4]. Electrospinning methods can be grouped into two: needle and needleless electrospinning. There are system and process parameters that have an effective role on the morphology of nanofibers. Process parameters can be defined as needle diameter, distance between electrodes, feed rate, applied voltage, ambient conditions while system parameters can be defined as polymer solution properties such as surface tension, viscosity, molecular weight, and solvent for the needle electrospinning system. Unlike the single needle electrospinning system, needleless electrospinning is useful equipment for significant bulk production in an industrial scale. Niu and Lin grouped the needleless electrospinning system as upward and downward [5]. One of the most popular upward needleless electrospinning systems is a roller electrospinning system that was developed by Jirsak et al. [6]. Despite many types of research about parameters of needle electrospinning, there is not enough information about needleless electrospinning.

The parameters of needleless roller electrospinning can be divided into two groups: dependent and independent. Each parameter has an influence on the resultant fiber morphology and spinnability. Cengiz and Jirsak [7] and Yener et al. [8] showed that the addition of tetraethylene ammonium bromide (TEAB) salt increases the spinning performance and resultant fiber diameter of polyurethane (PU) nanofibers. In another work, it was found that increasing the humidity of the spinning unit at a constant temperature, the spinning performance altered for polyvinyl alcohol (PVA) nanofibers while the nonfibrous area increases with humidity. On the other hand, at constant relative humidity, upon the increasing temperature from 20 to 30°C the spinning performance decreases almost three times so do the fiber diameter and nonfibrous area [9].

In this chapter, we focus on roller electrospinning and parameters. Independent parameters are the parameters that can be changed before or during spinning while dependent parameters are resultant parameters of independent ones. These parameters are explained in the next section. This work is a part of the Ph.D.thesis, the experimental and theoretical results were studied in detail in reference [10].

2. Principle of needleless electrospinning

Two common methods have been utilized for the production of nanofibers which are needle electrospinning and needleless electrospinning from a free surface. Both the spinning techniques were explained in the previous work [11]. The mechanism and the principle of the needle electrospinning system were explained deeply in the literature whereas there is not enough information about free surface needleless electrospinning.

Tonks [12] studied on the mechanism for the onset of aperiodic instability ($\omega^2 < 0$) of the free surface when subjected to a strong electric field. The initial perturbation of the surface causes a localfield increase which determines the further growth of the perturbation. The critical field is determined directly from the corresponding dispersion equation [13];

$$\omega^2 = \frac{k}{\rho}\left(\sigma k^2 + \rho g - \frac{E^2}{4\pi}k\right) \tag{1}$$

where g is the gravitational acceleration, σ is the surface tension, ρ is the density, E is electrical field strength, $k = 2\pi/\lambda$ is the wave number, and λ is the wavelength.

Lukas et al. explained the fasting-forming instability mechanism [14]. In the principle, the capillary waves on a one-dimensional approach of the liquid surface are used to analyze the electrohydrodynamics. The vertical displacement of the wave was calculated by the periodic real part of a complex quantity, $\xi = A\exp[i(kx{-}\omega t)]$. The A, k, and ω are amplitude, wave number, and angular frequency, respectively. The wave number (k) can be calculated using the wavelength (λ), $k = 2\pi/\lambda$. The relationship between the angular frequency ω and the period T is $\omega = 2\pi/T$ [10].

The relationships $\omega = f(k)$ between spatial and time-dependent parameters k and ω of waves undergoing various force fields that are called dispersion laws. For gravity waves, $\omega^2 = gk$, where g is gravity acceleration. By adding surface tension:

$$\omega^2 = (\rho g + \gamma k^2 - \varepsilon E_0{}^2 k)\frac{k}{\rho} \tag{2}$$

where ρ is liquid mass density, γ is a linear force of surface tension, k is wave number, ε is permittivity, and E_0 is field strength.

Lukas also gives out a new concept that is dimensionless electrospinning number Γ [15]. The dimensionless electrospinning number is the ratio of electric pressure (Pe) to capillary pressure (Pc). So we have, $\Gamma = Pe/Pc$. Capillary pressure (Pc) caused by an arbitrarily curved liquid surface is a product of the surface tension and a sum of two principal curvatures $K1$ and $K2$, so $Pc = \gamma(K1 + K2)$.

In the case of the sphere with a radius Rs in a gravitational acceleration field, both principal curvatures $K1$ and $K2$ are of the same value and are equal to $1/Rs$, so $Pc = 2\gamma/Rs$.

Electric pressure (Pe) is another basic concept in electrospinning, and it is calculated according to Eq. (3);

$$Pe = \left(\varepsilon E^2\right)/2 \tag{3}$$

where E is the strength of electric field and ε is the permittivity of surrounding air.

So, the phenomenon of electrospinning only occurs when the dimensionless electrospinning number is bigger or at least equal to 1 ($\Gamma \geq 1$ or $Pe \geq Pc$).

In the needle electrospinning, the polymer solution is drawn into fibers in electrostatic field arranged between a needle tip and a collector. Polymer solution stretches into fibers while the solvent is evaporating. Fibers in nanorange or microrange are collected on a collector. Fiber quality and morphology depend on several factors. The morphology of nanostructures formed on the collector varies with parameters of the spinning device and process parameters [10].

Dao et al. divided electrospinning processes into three stages: in the first stage, the polymer solution is fed by a roller from the container or by the pump. High electric potential inserted in the solution leads to the formation of jets. The jets (jet) are discharged from roller's surface Or droplet on the needle tip. As a result, the jets accelerate steadily and thin out along an axis aligned with the general direction of the electric field. It is a critical stage because the first stage is the key for the controlling subsequent stages and effects of the final fiber properties. In the second stage, several processes are forming simultaneously such as electric line fluctuation as a result of time and the space variation in the bulk density of charges causes jets to turn transversely to the field direction and to be decelerated by the constantly increasing drag force of the gas. A cloud is produced that expands toward the collector by the action of the same-polarity charges. At the same time, the rate of vaporization of the solvent which started already at the first stage of the process is steeply intensified; the jet solidifies, and the resulting fibrous cloud drifts in the applied electric field onto the collector. During this stage, an unsteady bulk fiber web structure can be observed due to possible splitting of the jet. In the third stage, there are two simultaneous processes: random deposition of the nanofibers onto a substrate on the collector and the gas spark discharge between the collector and the fiber layer formed on it, that closes the electric circuit [16].

3. Technology and parameters of roller electrospinning

The roller electrospinning has been developed and patented at the Technical University of Liberec [6]. The roller electrospinning system is a highly productive method to produce nanofibers. In this method, there is a roller electrode that is immersed in a solution tank and a motor connected to the roller for rotation. The roller electrospinning system is a closed system; there is only input for air and humidity and output for air suction. On the top there is a collector (charged or grounded), and a conveyor material (generally nonwoven or paper) is passing through the collector as shown in **Figure 1** [17]. Nevertheless, its productivity is still low in comparison with processes yielding planar fibrous materials, and the price of nanofiber layers is correspondingly high. Therefore, further development based on the knowledge of the process is necessary to improve quality, regularity, and price of

http://dx.doi.org/10.5772/65838

nanofibers. The roller electrospinning can process a broad range of polymers in diameters of 50–800 nm with acceptable narrow fiber diameter distribution on nonwoven webs.

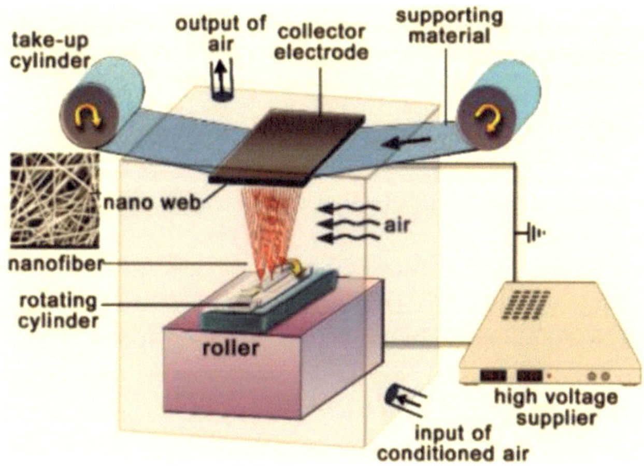

Figure 1. Schematic diagram of the roller electrospinning system.

Independent parameters	Dependent parameters
• Polymer solution properties (concentration, viscosity, composition, surface tension, conductivity, molecular weight) • Applied voltage • Distance between electrodes • Velocity or rotating roller • Velocity of take-up fabric • Geometry of electrode • Geometry of collector • Ambient conditions (temperature, relative humidity)	• Number of cones, density of jets • Lifetime of jets • Spinning performance, spinning performance/per jet • Total average current, current/jet • Thickness of polymer solution layer on the surface of the roller • Force acting on a jet • Spinning area • Distance between neighboring jets • Jet length in stable zone • Fiber diameter and distribution • Nonfibrous area • Launching time of jets

Table 1. Parameters of roller electrospinning system.

The electrospinning parameters are divided into two groups: namely independent and dependent ones. In the group independent parameters, primary and secondary independent parameters are distinguished, for instance kind of polymer, its molecular weight, kind of solvent, type and amount of additives, and concentration of polymer solution are primarily independent parameters, whereas viscosity, conductivity, and surface tension, as a result of above parameters, directly influencing the electrospinning process, are secondary independent parameters [10].

The parameters of the roller electrospinning system were tabulated in **Table 1** [10].

3.1. Independent parameters

3.1.1. Polymer solution properties

3.1.1.1. Concentration/molecular weight/viscosity

Polymer jets must show some level of strength to create nanofibers during electrospinning. If the strength is too low, the jets break and create beads instead of fibers. Shenoy et al. [18] introduced the term "entanglement number" as a quantity responsible for the strength of jets. They also found the limiting value of an entanglement number regarding fiber production; if the value is greater than 3.5, fibers are produced, below 3.5 the production of beads starts. The entanglement number depends on the kind of polymer, namely on the content of polar groups and molecular weight, on the type of solvent and polymer concentration in the solution. As the determination of the entanglement number is not easy, the viscosity is often taken as a measure of it, as its principle depends on the same parameters as entanglement number. In a suitable concentration, the chain entanglements of macromolecules are adequate to electrospin a polymer into fibers under a strong electrostatic field.

The polymer solution viscosity is determined by such molecular interactions, either attractive or repulsive, and can be altered by changing the solvent type. The viscosity of polymer solutions depends on shear rate and temperature, it also depends on the molecular weight distribution and the concentration. Properties of solvents are affecting the chain entanglement. In a good solvent, the polymer swells and its volume increases. The intermolecular forces between the monomer and solvent subunits dominate over intramolecular interactions. In a bad solvent or poor solvent, intramolecular forces dominate and the chains contract. In theta-solvent, the intermolecular polymer-solvent repulsion balances exactly the intramolecular monomer-monomer attraction [19]. In aqueous, the overall shape of the chain will be affected by the amount of electrolyte (salt) in the solvent. In the absence of electrolyte, the charges are unshielded and repel each other. As a result, the chain is stretched out. However, when the electrolyte is present in the solution, the charges are shielded, and their effect is suppressed. As a result, the chain tends to shrink toward its original random configuration. This change from a rod to a sphere obviously causes the viscosity to decrease [20].

With the increasing concentration of polymer solutions prepared by using polymers with different molecular weights, the viscosity of polymer solutions was found to increase due to the densely entangled polymer chains [21]. A certain amount of chain entanglement is needed to keep the solution jet coherent during the electrospinning process. The high viscosity or the concentration of the solution prevents the jet to breakup into droplets. Moreover, the increase in chain entanglements results in an increase of viscoelastic force, which can be enough to prevent breakup of the electrically charged jet. As a result, the charged jet elongates through a collector by Coulombic stress and the diameter of jet decreases [22]. With any further increase in the concentration (or viscosity) of the solution, the number of beads along the fibers was found to decrease, and their shape appeared to be more elongated [23]. The relationship between polyvinyl butyral nanofiber surface morphology, molecular weight and concentration are shown in **Figure 2**.

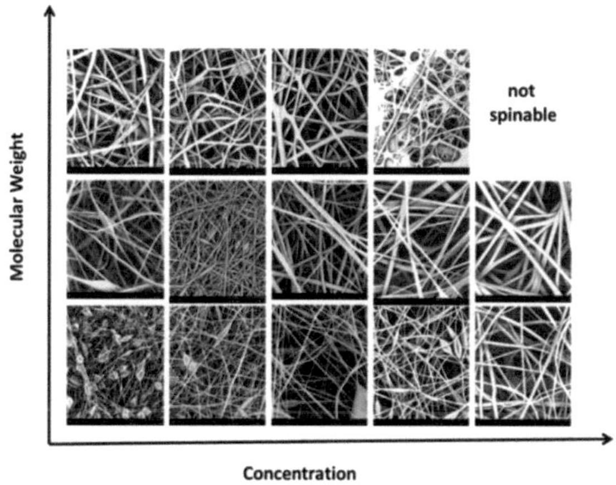

Figure 2. The relationship between the surface morphology of nanofibers, molecular weight, and concentration.

Cengiz et al. [24] studied the effect of PVApolymer molecular weight and some solution properties such as conductivity, surface tension, and rheological behaviorconcerning spinnability, process performance, fiber diameters, diameter distribution, and nonfibrous particles. It was found that electric conductivity and surface tension of the solutions did not influence both spinning performance and fiber diameter significantly, whereas molecular weight has an important effect on the spinnability. On the contrary, the concentration changes the process throughput considerably and properties of nanofibers and nanofiber layers to some extent.

In general, the relation between molecular weight, viscosity, and concentration, the number of entanglements per chain increases with molecular weight while entanglement per unit volume increases with concentration. In the previous studies, it was found that increasing concentration increases viscosity, and as a result, more polymer solution is transferred per one jet. Therefore, fabric throughput and fiber diameter increase [7, 16].

3.1.1.2. Surface tension

Surface tension tends to decrease the surface area per unit mass (and the surface energy) of fluid. If the electrostatic force overcomes the surface tension, a charged jet of a polymer solution is ejected. So, for a higher surface tension, a stronger electric field is required. Surface tension is related to the interaction between the molecules of polymer and solvent in solutions.

Fong et al. found that the number of beads decreased with increasing viscosity and net charge density and it also decreased with decreasing surface tension coefficient of the solutions [25].

Surface tension causes the solvent molecules to form spherical shape and aggregate when the free solvent molecules are in high concentration. The interaction between the polymer and the solvent molecules is high at high viscosity and results in the elongation of the solution under the electrical field. As a result of the electrical field, the solvent molecules will spread out from the polymer molecules and reduces the gathering of the solvent molecules under the effect of the surface tension.

Figure 3 shows the PVA solution at various surface tension resulting in changes in fiber surface morphology. Nonionic surfactant was used not to change conductivity.

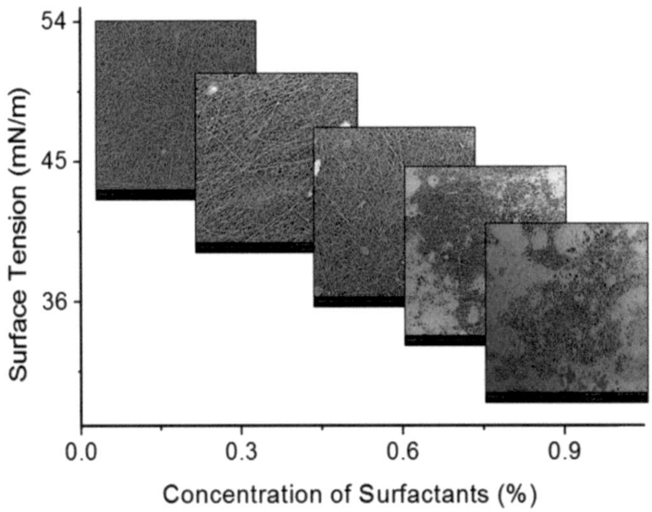

Figure 3. Effect of surface tension on the surface morphology of nanofibers.

3.1.1.3. Conductivity and permittivity

The electrospinning process operates due to the external electric field, the viscoelastic force due to the viscosity of the solution, surface tension force, conductivity, etc. The electric field between the electrode and the collector causes the solution to stretch. If the electrostatic field overcomes the surface tension, the nanofibers are formed. The stretching of the electrospun jet and the bending instability is mainly controlled by the Coulombic force between charges and the electric field. These forces arise due to the surface charge on the jet which can be varied by changing the conductivity of the solution. The increased charge carrier causes the jet to elongate and can produce uniform fibers with a smaller diameter. A higher net charge density of the polymer solution could, therefore, yield thinner fibers with no beads. Filatov et al. found that the acceptable range of conductivity in electrospinning is rather wide from 10^{-6} to 10^{-2} $(\Omega\ m)^{-1}$, where the upper limit is governed by the threshold of initiation of the gas discharge from the jet that destabilizes it. It was also discussed that the conductivity is associated with relative dielectric permittivity which usually varies little from its value of the solvent that is used. The lower the permittivity, the lesser the attenuation of the electric field inside the nascent liquid jet and the higher the rate of electric charge transfer in the solution. However, decreasing of the polarity of solvent molecule reduces the extent of dissociation in it of ionogenic substances with a concurrent reduction in electrical conductivity. For the electrospinning process, the best values of relative permittivity were found to be between 5 and 30, with a limit not exceeding 100 [26]. The conductivity of the solution is the factor that determines the current and fiber diameter. The solution with high conductivity can carry more charge on the jet. Kim et al. found that nanofibers with the smallest fiber diameter can be obtained from the solution with the highest conductivity [27]. On the other side, Heikkila and Harlin found that at the same viscosity, polyacrylonitrile (PAN)/salt solution produced slightly larger fibers because increased conductivity enhanced the mass flow. The higher conductivity of the PAN/salt solution increased the instabilities in the electrospinning process [28].

It is common to use salt in polymer solutions to increase their conductivity. According to a previous work, the addition of salt to polymer solution such as polyurethane and polyvinyl butyral (PVB)enhances the number of jets on the roller surface and as a result the productivity increases [7, 29]. On the contrary, in the case of some other polymers such as polyethylene oxide (PEO) and PVA, the addition of salt decreases the spinning performance [30, 31]. This case can be explained by the leaky dielectric model that was first proposed by Melcher and Taylor [32].

3.1.2. Applied voltage

It was investigated that a drop of liquid having surface tension γ, held at the end of a capillary tube of radius r_0, and raised to a potential difference of V, disintegrates into a spray and at the point of disintegration the ratio $V^2/r_0\gamma$ remained approximately constant [33]. Then Taylor [34] found that the critical angle of the meniscus which is called as "Taylor's cone" closed to disintegration is 49.3°. On the other hand Yarin et al. showed that Taylor's cone represents a specific self-similar solution while nonself-similar solutions that do not tend to the semivertical angle. The shape of the meniscus changes as the concentration of the polymer in solution and the solution viscosity increases [35]. Another work showed that conductivity of solution changes the shape of the meniscus. Various voltages have significant effects on droplet size, fiber diameter, and current transport. It has been verified experimentally that the shape of the initiating drop changes with applied voltage [36].

The shape of the surface changes significantly under a critical voltage. The surface changes and the electrospinning current, which significantly depends on voltage, increase in the bead formation, number of beads, and the defects on the surface of electrospun fiber web [37].

Lukas et al. analyzed the self-organization of the liquid jet [14]. In this analysis, it is supposed that electrohydrodynamics of a liquid surface can be analyzed with the capillary waves running on a one-dimensional approximation of the liquid surface, oriented along the horizontal axis, like *x-axis*, of a Cartesian system of coordinates. The critical electric strength E_c for unstable waves that start to form cone was formulated as:

$$E_c = \sqrt[4]{\frac{4\gamma\rho g}{\varepsilon 2}} \tag{4}$$

where γ is the linear force of surface tension, ρ is the mass density of liquid, g is the gravitational acceleration, ε is the permittivity. The above assumption indicated that the critical electrical field strength, E_c, allows the creation of surface wave for initiating of jets.

From the previous work [31] it was found that upon increasing applied voltage the number of cones increases on the surface. As a result, fabric throughput increased. All these assumptions above are for needleless electrospinning. The minimum electric field depends on the surface tension and density of the solution, relative permittivity of the surrounding gas, and gravitational acceleration. As soon as the electric field increases and exceeds the critical electric field, a charged jet ejects from the apex of the cone and deposits on the collector.

3.1.3. Distance between electrodes

The distance between electrodes is another phenomenon that controls the final fiber diameter and morphology. Sufficient time to dry out the solvent from a polymer solution is necessary. If the distance between electrodes is too short, there is not enough time for drying, beaded, and sticky fiber structures can be observed. On the contrary, if the distance is too high, the electric field between electrodes decrease and forming of fiber is difficult.

Yuan et al. found that there were no obvious differences between the morphologies of the electrospun polysulfone (PSF) fibers at the 10–15 cm electrode to electrode distance. Increasing the collection distance could give more time for the solvent to evaporate and for the charged fluids to split more times [38]. The effect of the tip to collector distance on the fiber morphology is not as significant as that of other parameters, and this has been observed by many investigators [39–41].

3.1.4. Ambient conditions

Electrospinning is affected by an electric field. It means that any change in the electrospinning environment will affect the electrospinning process. When humidity is high, it is likely that water condenses on the surface of the fiber. At a very low humidity, a volatile solvent may dry out very rapidly. The evaporation of the solvent may be very fast. As a result, the electrospinning process may only be carried out for a few minutes before the needle tip is clogged. Many studies showed that ambient conditions (relative humidity, temperature) influence fiber diameter, morphology, and spinning performance [12].

3.1.5. Velocity of rotating roller

The main function of the roller consists in feeding the polymer solution from a tank in the form of solution layer on the roller surface into the electrospinning process between the spinning and collector electrode. The role of velocity of the rotating roller can be summarized as:

- Arrangement of the thickness of polymer solution layer on the surface of roller.

http://dx.doi.org/10.5772/65838

- Changing the feed rate of polymer solution [42] (**Figure 4**).

- The time available for formation Taylor's cones and the spinning process from these cones.

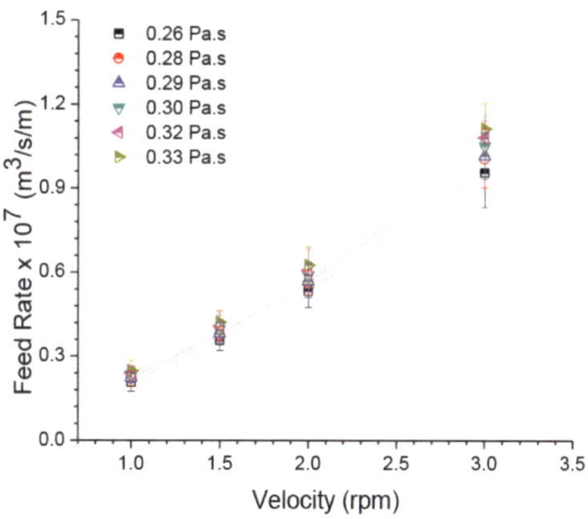

Figure 4. Feed rate of various PEO solution layers as a function of roller rpm and viscosity.

The speed of roller significantly influences the quality of the resultant nanofiber web, as shown in **Figure 5**:

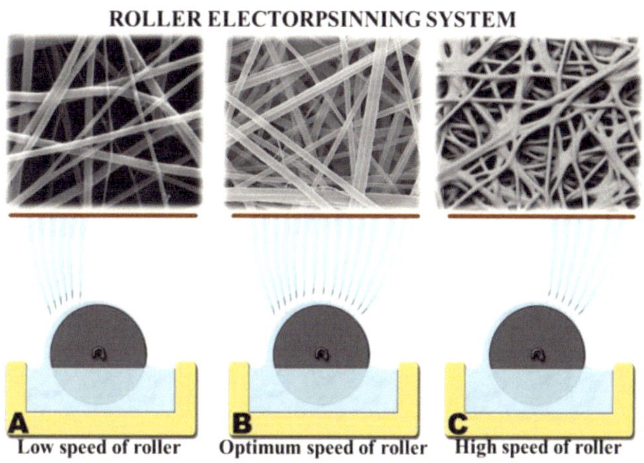

Figure 5. Effect of roller speeds on the quality of nanofiber web.

At the very low speed of roller (A), there will be not enough solution on the roller surface, so the number of cones forms on the roller surface and spinning performance will be very low. It is necessary to arrange roller speed as optimum (B) to get the best quality of nanofiber layer and high throughput. If the roller speed is very high (C), there might be a significant amount of solution supplied to the top of the roller, but there will be not enough time to form Taylor 's cones on the surface of the roller. The fiber quality will decrease.

3.1.6. Velocity of take-up fabric

During the roller electrospinning process, a supporting material (often a nonwoven fabric) is passing along the collector electrode to collect the fibers in the form of a more or less regular nanofiber layer on its surface. The speed of this fabric (in meters per minute) influences the area weight of a nanofiber layer and also affects the quality of nanofiber membrane, namely its regularity and nonfibrous area. The area weight of the resultant nanofiber layer can be easily calculated from the spinning performance and the velocity of supporting materials. Typical values of the velocity amount from 0.1 to 10 m/min.

3.1.7. Geometry and conductivity of the collector electrode

In the needleless ES, many jets are formed simultaneously on the surface of the spinning electrode. Jets are generated on the free liquid surface by a self-organizing process. Because of this, it is harder to control the spinning process when compared with the needle electrospinning process. The spinnerets in the needle electrospinning play an essential role in determining the product quality and productivity. Despite the success in needleless electrospinning, a design of a principle for a better needleless electrospinning process is still in progress. That is why the effort of scientists was focused on the shape of spinning electrodes.

Niu et al. [43] studied the influence of spinner shape on the electrospinning process using three rotating fiber generators, cylinder, disc, and coil. The electric field generated in the vicinity of these spinners was analyzed by a finite element method, and the calculations indicated that electric field intensities on these spinnerets are distributed differently. It was observed that, on the cylinder spinneret, the electrical field was distributed unevenly. The middle part of the cylinder spinneret had lower strength. Among all the spinneret, only the coil surface showed an intensity peak. In comparison of the cylinder and the needle electrospinning system, finer nanofibers with narrower fiber diameter distribution were produced by using coil and the disc spinneret. Compared to needle, cylinder, and disc electrospinning, higher productivity was achieved as 23 g/h using coil electrospinning.

The collector electrode in electrospinning is a conductive material. The collector can be directly connected to a high voltage supplier that gives the opposite charge than the spinning electrode has, or it can be grounded. Fibers that are collected on the nonconducting material usually have a lower packing density compared to those collected on a conductive surface.

3.2. Dependent parameters

3.2.1. Number of cones

Some Taylor's cones are simultaneously present on the spinning surface of the spinning roller electrode. Some cones (N_c) depend strongly on independent electrospinning parameters.

Cones were counted using a camera record as shown in **Figure 6** and their number is related to the spinning area determined from the same record [10].

The average number of cones (N_c) is counted from these pictures. The density of cones (D) can be calculated as the ratio of number of cones (N) to the spinning area (A).

$$D = N_c / A \left(1/m^2\right) \tag{5}$$

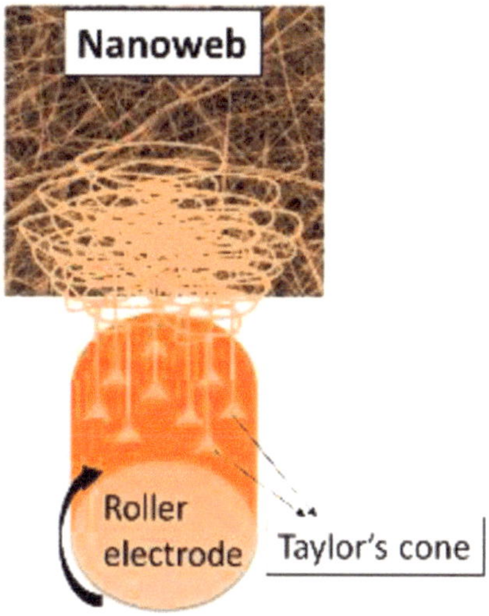

Figure 6. Number of cones on the roller surface.

3.2.2. Spinning performance and spinning performance per jet

Spinning performance (SP) is one of the most important characteristics of the nanofiber production process, influencing the production costs. The spinning performance of a single needle electrospinning is quite low while rather a high performance can be achieved by the needleless system. Spinning performance strongly depends on all the independent parameters and influences the quality of nanofibers as well as that of nanofiber layers.

SP can be determined from the mass of nanofibers produced in a 1 m long roller spinning electrode in 1 min. Spinning performance is calculated from area weight of produced nanofiber layer as follows:

$$\mathrm{SP} = \frac{G v L_{\mathrm{f}}}{L_{\mathrm{r}}} [\mathrm{g/min/m}] \tag{6}$$

where G is the area weight of nanofibers membrane per area in g/m^2; v is the velocity of running collected fabric, in m/min; L_f is the width of nanofibers membrane on collected fabric, in m; and L_r is the length of spinning roller, in m.

Spinning performance per one Taylor's cone (SPC) can be calculated from the known values of SP and an average total number of Taylor's cones on the spinning electrode N_c using the camera images.

3.2.3. Fiber diameter and fiber diameter distribution

The fiber diameters are easily measured on the scanning electron microscope (SEM) microphotographs using software.

3.2.4. Nonfibrous area

The term nonfibrous area (**Figure 7**) was created as a measure of nanofiber layer quality which is a percentage of the area of an SEM microphotograph of nanofiber layer occupied by nonfibrous formations such as beads, foils, and so on.

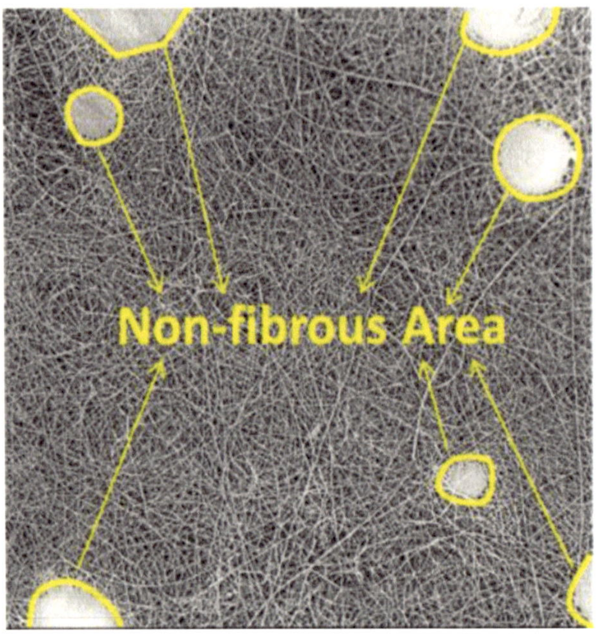

Figure 7. Nonfibrous area on the surface of nanofiber web.

3.2.5. Lifetime of a jet

A lifetime of a jet can be defined as the period from the point that the jet appears on the spinning electrode to the point that the jet disappears. The solution properties, roller speed, electric field strength, the number of cones, and the thickness of a layer on the roller surface are the main parameters that affect the lifetime of the jet. The lifetime of a jet can be measured using a camera

http://dx.doi.org/10.5772/65838

record of the spinning area during electrospinning. The measurement procedure consists in tracing a specific Taylor's cone during step-by-step playing the record.

3.2.6. *Average current and current per jet*

Under an electrostatic field, the polymer solution droplet becomes distorted by the induced electrical charge on the liquid surface, and a stable jet of the polymer solution is then ejected from the cone. The breakup of the jet depends on the magnitude of the applied electric current [10]. The bending instability in a charged jet was characterized by Reneker and Chun during electrospinning [44]. Many researchers on electrospraying have focused on the understanding of a current-voltage relationship. Previous attempts to correlate current with the electric field and flow rate suggested nonunique linear [45] or power law [46] dependencies. Using the power law, Fallahi et al. [47] showed that the relationship between current and voltage as current ~ $(\text{voltage})^{2.53}$. Demir et al. [48] reported a power law relationship between the solution flow rate (Q) and the applied voltage (V) with an exponent value of three in the experiments where Q was not controlled. Bhattacharjee et al. [49] showed that current depends not only on applied voltage but also on conductivity and feed rate of the polymer solution. The relationship between current and other parameters was indicated that [49]

$$I \sim E^{*} Q^{0.5^{*}} K^{0.4} \tag{7}$$

Theron et al. [46] reported that the average, electric current increased with the increase in the feed rate and voltage. It was also found that with increasing the applied voltage and solution conductivity both the jet current and fiber diameter were increased [47]. Because increasing the electric field increases the electrostatic stresses, which in turn, draws more material toward the collector. Kim et al. found that there is an optimum average electric current for the electrospinning process that depends on the solution properties [50]. Measurement of current per one jet was suggested in the previous work [13]. The strong relationship between the current of polymer solution jet and spinnability as well as spinning performance was demonstrated by Cengiz et al. using a rod electrospinning system. They found that electric current of the solution jet increases with the increasing concentrations of PU and TEAB, and current indicates the spinning performance [51].

3.2.7. *Thickness of polymer solution layer on the surface of the roller*

The effect of roller rotation on the thickness of polymer solution layer has been examined in several papers [52–54]. There have been several theoretical and experimental advances. Frenkel et al. investigated the nonlinear stability of a film flowing down a cylinder using the strong surface tension approximation and showed that wall curvature significantly influences the transverse disturbance of the film, suppressing harmonics of high angular frequency [53]. In a previous work, it was suggested the measurement of the thickness of layer on the surface of rotating roller using a custom-built micrometer, as shown in **Figure 8** [42]:

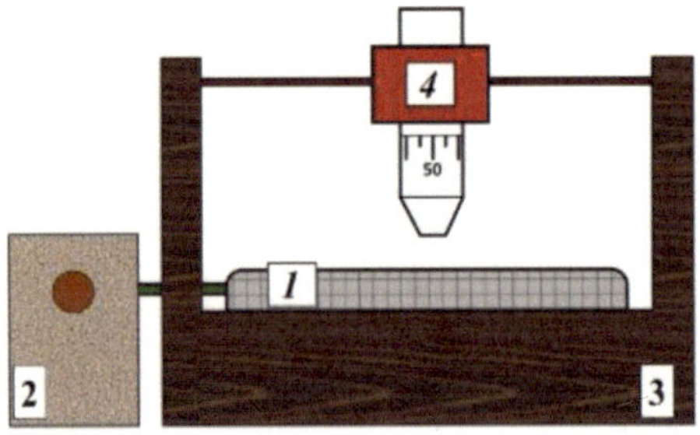

Figure 8. Number (1) is the rotating roller with a diameter of 20.0 mm driven by an electrical motor; (2) and (3) are the polymer solution tank; (4) is the micrometer.

The relationship between thicknesses of layers on the roller surface and roller speed is found proportional [42], as shown in **Figure 9**:

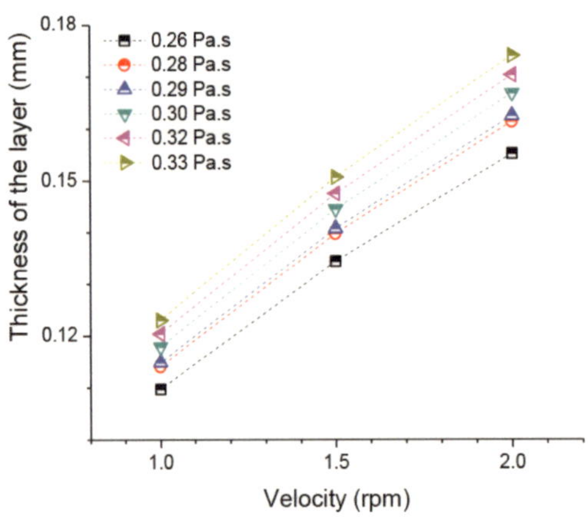

Figure 9. The thickness of various PEO solution layers as a function of roller rpm and viscosity.

3.2.8. Force acting on a jet

The analysis and calculation of the force acting on a jet versus flow of polymer solution into this jet were performed during the spinning process. The Navier-Stoke equation model has been proposed to describe the force. Relations between the force acting on a jet and the volume flow rate $\dot{V}$, dynamic viscosity μ and thickness of the fluid layer h were calculated in the literature [10, 55].

It seems to be tough to calculate the force acting on the jet based on the Coulomb's law as the electric field between the electrodes in the electrospinning device is certainly not static. Therefore, it was calculated the force from the volume flow rate, the thickness of polymer solution layer on the roller spinning electrode and the diameter of the base of Taylor's cone.

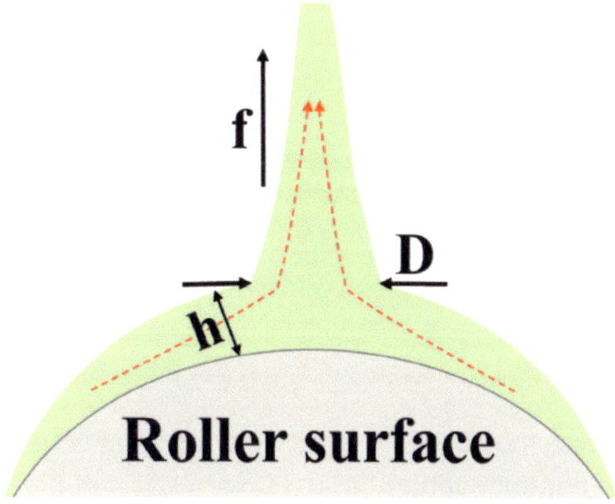

Figure 10. Schematic diagram of Taylor'scone. f—electric force, h—thickness of the fluid layer, D—diameter of the base of Taylor's cone.

The calculation relies on the idea that the polymer solution flows from the layer on the spinning roller into the jet through the circular area of radius R and the height h as shown in **Figure 10**. The flow is described using the Navier-Stokes equation as follows:

3.2.8.1. Steady laminar stokes flow over a plate

Full formulation of the momentum equations of the Navier-Stokes equation is:

$$\frac{\partial \mathbf{u}}{\partial \mathrm{t}} + \frac{\partial (\mathbf{uu})}{\partial \mathbf{x}} = \frac{\partial p}{\partial \mathbf{x}} + \mu \frac{\partial^2 \mathbf{u}}{\partial \mathbf{x}^2} + \boldsymbol{f} \tag{8}$$

Eq. (8) is rewritten in the form only for relevant direction x and the first unsteady term is zero, the second term depicts nonlinear flow behavior, and it is zero, the pressure gradient is zero, the viscous term is partially neglected, external force remains and so the simplified equation after revision is as follows:

$$\mu \frac{\partial^2 u}{\partial y^2} = -f_y = -f \tag{9}$$

The next step is to find the velocity as a function of the *y-coordinate,* so Eq. (9) must be integrated and coefficients will be determined from the boundary conditions (b.c.) as follows:

$$\frac{\partial^2 u}{\partial y^2} = -\frac{f}{\mu} \tag{10}$$

$$\frac{\partial}{\partial y}\left(\frac{\partial u}{\partial y}\right)=-\frac{f}{\mu} \tag{11}$$

$$\left(\frac{\partial u}{\partial y}\right)=-\frac{f}{\mu}y+C_1\cdots\ b.c.:\cdots b.c.:y=h,\tau=\mu\frac{\partial u}{\partial y}=0,\cdots\frac{\partial u}{\partial y}=0C_1=\frac{f}{\mu}h \tag{12}$$

Note: the previous condition took account of the Newton fluid behavior. This place should be revised in the case of the non-Newtonian fluid. However, if non-Newtonian fluid will be still based on the velocity gradient, then this relation will be valid for non-Newton fluids as well.

$$\left(\frac{\partial u}{\partial y}\right)=-\frac{f}{\mu}y+\frac{f}{\mu}h \tag{13}$$

$$u(y)=-\frac{1}{2}\frac{f}{\mu}y^2+\frac{f}{\mu}hy+C_2 \tag{14}$$

$$\text{b.c.}: \text{y}=0,u=0,\ldots.C_2=0 \tag{15}$$

The velocity depends on the *y*-coordinate:

$$u(y)=-\frac{1}{2}\frac{f}{\mu}y^2+\frac{f}{\mu}hy \tag{16}$$

After formal rewritten:

$$u(y)=\frac{1}{2}\frac{f}{\mu}\left[2hy-y^2\right] \tag{17}$$

Eq. (17) describes the velocity distribution on the *y*-coordinate so-called the velocity profile.

Mass flow rate:

$$\dot{Q}=\int_S u(y)\rho dS=\frac{1}{2}\pi D\rho\int_0^h\frac{f}{\mu}\left[2hy-y^2\right] \tag{18}$$

$$dy = \frac{\pi Df}{2\vartheta}\left[hy^2 - \frac{1}{3}y^3\right]_0^h = \frac{\pi Dfh^3}{3\vartheta} \tag{19}$$

where ϑ represents a kinematic viscosity in units m^2/s.

External local force per volume of the fluid is calculated from Eq. (19) and takes a form as follows,

$$f = \frac{3\dot{Q}\vartheta}{\pi Dh^3} = \frac{3\dot{V}\rho\mu}{\pi Dh^3\rho} = \frac{3\dot{V}\mu}{\pi Dh^3} \tag{20}$$

where $\dot{Q}$ is the mass flow rate (kg/s), $\dot{V}$ is the volume flow rate (m^3/s), ϑ is kinematic viscosity, and μ is dynamic viscosity, respectively.

The total force for the whole volume of the fluid in a Taylor cone is:

$$F = \int_V fdV = \int_0^{2\pi}\int_0^h\int_0^R frd\varphi dydr = \frac{3\dot{V}\mu}{2\pi h^3}\int_0^{2\pi}\int_0^h\int_0^R \frac{1}{r}rd\varphi dydr = \frac{3V\dot{\mu}R}{h^2}\left[\frac{\mathbf{kgm}}{\mathbf{s}^2}\right] \tag{21}$$

In fact, the force F depends on electrical characteristics of the space between electrodes and on applied voltage. The quantities on the right side of Eq. (21) influence each other in such a manner that Eq. (21) remains true when any of these quantitiesis changed.

For instance, the increase in the thickness h causes a corresponding increase of V and/or R.

The influence of the fluid viscosity seems to be out of validity of Eq. (21). The growth of the solution viscosity depending on both polymer concentration and its molecular weight is linked with a greater degree of macromolecules entanglement and with the strength of jets that increases the volume flow rate. Above-mentioned relations make it rather difficult to study influences of specific parameters separately.

The results are a theoretical base for future experimental investigations. Nevertheless, some specific influences must be taken into account in planning and evaluation of the experiments which are discussed above [10, 55].

3.2.9. Spinning area and the position of the jets

The spinning area is the place where the Taylor's cones occur on the roller surface. The experimental results showed that the spinning area mainly depends on the parameters mentioned in **Table 1** such as speed of roller, polymer solution properties, electrical field, additives used in solution, and number of jets [10, 42]. The relationship between the spinning area and the position of jets was explained in Section 3.1.5 (**Figure 5**).

The thickness of a polymer solution layer on the roller influences the solution to flow into Taylor cones. Simultaneously, polymer solution, when moving out of the tank, needs a certain time to create Taylor's cones. Subsequently, the solution from the vicinity of Taylor's cones is dragged into the cones and following jets, being consumed gradually. If the velocity of the roller is slow, there is only a small amount of solution brought to the surface, and the solution is spun away before it reaches tank again. Thus, only a part of the potential spinning area is utilized. On the other hand, if roller velocity is too high, there is not enough time for a solution to create Taylor's cones and the spinning area is narrow again. The spinning area can be determined by using a camera.

3.2.10. Distance between neighboring jets

The average interjet distance is Describe dregarding wavelength (λ) as Lukáš et al. calculated it [15]. According to this formula, electric field strength, surface tension, density of the solution, the relative permittivity of the surrounding gas, and gravitational acceleration are the main parameters that affect the interjet distance [10]. In other studies [11, 42], the distance between neighboring jets was calculated using the number of jets (n) and spinning area (A). The spinning area A is divided into n identical squares belonging to Taylor's cones.

$$A = n.x^2 \tag{22}$$

$x(m)$ is the length of the square side (and simultaneously the distance between Taylor cones). Then the average distance x between adjoining Taylor cones is

$$x = \sqrt{\frac{A}{n}}\,(\mathrm{m}) \tag{23}$$

3.2.11. Length of the jet in stable zone

The length of the jet of the polymer solution was described and studied by Dao [16]. The distance between the tip of a Taylor cone and the splitting point of a jet is called length of the jet. The long-range repulsive electrostatic forces between ions of the same sign cause the disintegration of charged jet body. If electric forces are stronger than the capillary forces, the jet will split or will be formed by the whipping instability. The process of concentration of ions at the surface of a jet preceding the onset of splitting or whipping instability is shown in **Figure 11** [10]. The condition of the process, such as solution viscosity, the amount of ions per volume unit, etc., predetermines the length of a jet.

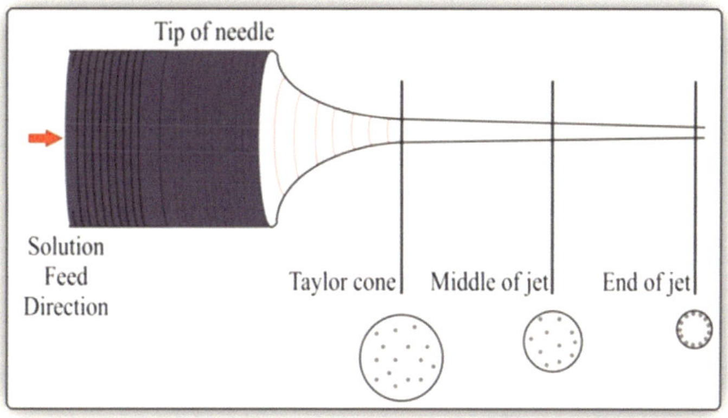

Figure 11. Radial movement of ions in a jet.

In the polymer solution entering the electrospinning process, the ions are more or less regularly distributed in the volume. When the stretching of a jet starts, the ions move toward the surface of the jet due to repulsive forces. As soon as the concentration of ions at the surface and the corresponding repulsion forces are big enough, the disintegration process starts. The viscosity and conductivity of solution are the key parameters influencing the time necessary for the onset of disintegration. If the viscosity is high, movement of ions toward the surface takes more time,and the length of the jet will be greater [16]. The addition of salt increases the conductivity of solution as well as a number of ions in the solution. If a number of ions are high, the time for ions to reach critical concentration will be shorter. Consequently, the jet will be shorter [16].

3.2.12. Launching time of jet

Highly concentrated solutions have an intensive entanglement of polymer chains. Therefore, coulombic repulsion force will not be sufficient to start jetting. However, the surface area has to be increased to accommodate the charge build-up on the jet surface which occurs through the formation of fibers. Deitzel et al. [56] suggested mobility of particles in electrospinning jet initiation to occur from the surface layers of the cone. The reason is partly due to surface shear forces generated by the potential difference between the electrodes.

The time needed for Taylor cone formation (characteristic hydrodynamic time) was experimentally determined from voltage-time oscilloscope records of the electrospinning onset event that were accompanied by video-records [57]. It was hypothesized that the characteristic hydrodynamic time is of the same order as the theoretically derived reciprocal value of the wave growth rate that is denoted here as "relaxation time." The derivation was based on the linear stability analysis. It was found that these times are of dynamic nature and hence they are viscosity dependent. The theoretical derivation of relaxation time for conductive viscous liquids was introduced and related to experimental measurements of characteristic hydrodynamic times. It was assumed that relaxation time T is the reciprocal quantity to the growth factor A, $A = 1/T$.

$$A = -O_h K^2 + \sqrt{O_h^2 K^4 - \Omega_0^2} \tag{24}$$

where Ω_0^2 is the dimensionless square of the angular frequency of a nonviscose liquid, A is the dimensionless growth factor of a viscose liquid; K is the dimensionless wavenumber and O_h is the Ohnesorge number.

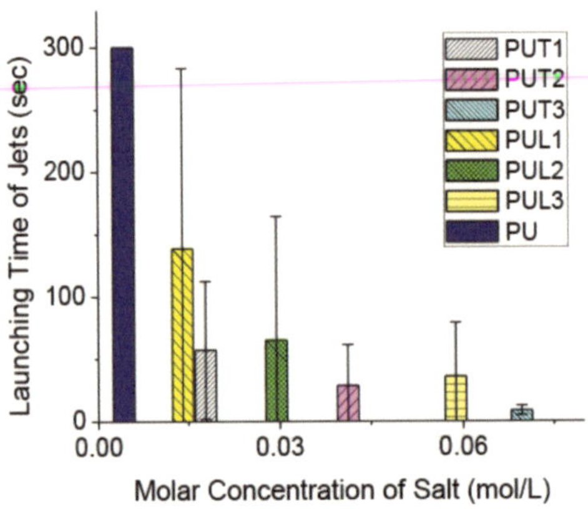

Figure 12. Launching time of jets according to the various molar ratio of tetra ethylene ammonium bromide (T) and lithium chloride (L) content of salts in polyurethane (PU) solution.

Recently, it was suggested the measurement of launching time by observing and recording the timing. The results indicated that additives have asignificant influence on the launching of jets as shown in **Figure 12** [10].

Additional salt increases the conductivity of the solution. The charged solution can easily influence from the electrical field between the electrodes and yields to form Taylor's cone.

4. Conclusion

A needleless electrospinning system is a significant method for the industrial production. Compared to the needle electrospinning system, there still requires more work to define the parameters and measurement methods of parameters. This chapter summarizes the parameters of needleless electrospinning. First, parameters grouped as dependent and independent parameters. Then, the relationship between parameters is defined, and the measurement methods were introduced systematically. The results can be concluded that:

- The total electric current in the roller electrospinning is more or less proportional to the spinning performance. On the contrary, the current per jet is not commensurate with the spinning performance per jet. This discrepancy evoked a more thorough investigation of the phenomena. The results were separately published [11] and are not a part of this chapter;

- The relationship between the number of jets and the spinning performance on the roller velocity is generally nonmonotonous. The main influence factors for the relation are the thickness of solution layer and feed rate versus movability of jets against the roller surface [10, 42];
- The launching time of jets does not influence the number of jets on the spinning roller and spinning performance significantly. It only can affect the starting period of the spinning process. As soon as the spinning process becomes stable, the number of jets depends on other parameters and is principally stable;
- The experimentally found distance between jets is in a reasonably good agreement with theoretical conclusions of Lukas. Nevertheless, it depends on some more variables than Lukas considered, namely on the conductivity of polymer solution;
- The spinning area mainly depends on the movement of the roller. Experimental results of PEO showed that increasing roller rpm decreases the spinning area except PEO without salt. The relationship between the spinning area and spinning performance is related to the number of jets. The higher number of jets means the larger spinning area which also means higher SP [42];
- The length of jets mainly depends on the amount of salt which is related to conductivity and the viscosity of the solution. Experiments showed that there is no relation between SP and length of jets [10];
- The effect of the rotating roller on fiber surface morphology indicates that increasing roller speed increase the fiber diameter and nonfibrous area. It is necessary to optimize roller speed to obtain proper fiber diameter and minimum nonfibrous area for every polymer solution [10, 42];
- The salts may influence the entanglement number and polarity of macromolecules when creating complex bonds with them. On the other hand, the salts increase the conductivity of solutions which may cross the limits supposed by the model of a "leaky dielectric."

The role of all parameters is important for better productivity of nanofibers in fine diameter.

Author details

Fatma Yalcinkaya[1*], Baturalp Yalcinkaya[1] and Oldrich Jirsak[2]

*Address all correspondence to: yenertex@hotmail.com

1 Institute for Nanomaterials, Advanced Technology, and Innovation, Technical University of Liberec, Liberec, Czech Republic

2 Department of Nonwovens and Nanofibrous Materials, Faculty of Textile Engineering, Technical University of Liberec, Liberec, Czech Republic

References

[1] Vasita, R. and D.S. Katti, *Nanofibers and their applications in tissue engineering*. International Journal of Nanomedicine, 2006. 1(1): pp. 15–30.

[2] Xue, J.J., et al., *Drug loaded homogeneous electrospun PCL/gelatin hybrid nanofiber structures for anti-infective tissue regeneration membranes*. Biomaterials, 2014. 35(34): pp. 9395–9405.

[3] Heo, D.N., et al., *Burn-wound healing effect of gelatin/polyurethane nanofiber scaffold containing silver-sulfadiazine*. Journal of Biomedical Nanotechnology, 2013. 9(3): pp. 511–515.

[4] Yalcin, I., et al., *Design of polycaprolactone vascular grafts*. Journal of Industrial Textiles, 2016. 45(5): pp. 813–833.

[5] Niu, H.T. and T. Lin, *Fiber generators in needleless electrospinning*. Journal of Nanomaterials, 2012. 2012: p. 13.

[6] Jirsak, O., et al., *A Method of Nanofibers Production from a Polymer Solution Using Electrostatic Spinning and a Device for Carrying Out the Method*, in *EP1673493*. 2005: Czech Republic.

[7] Cengiz, F. and O. Jirsak, *The effect of salt on the roller electrospinning of polyurethane nanofibers*. Fibers and Polymers, 2009. 10(2): pp. 177–184.

[8] Yener, F., et al., *New measurement methods for studying of mechanism of roller electrospinning*. Nanocon 2013, 5th International Conference, 5th–7th November 2014, Czech Republic: pp. 808–813.

[9] Dao, A.T., O. Jirsak, and T. Ltd, *Roller electrospinning in various ambient parameters*. Nanocon 2010, 2nd International Conference, 2010: pp. 103–108.

[10] Yener, F., *New methods in the study of roller electrospinning mechanism*, in *Department of Nonwoven and Nanofibrous Materials*. 2014, Technical University of Liberec: Liberec. p. 136.

[11] Yener, F., B. Yalcinkaya, and O. Jirsak, *On the Measured current in needle- and needleless electrospinning*. Journal of Nanoscience and Nanotechnology, 2013. 13(7): pp. 4672–4679.

[12] Tonks, L., *A Theory of liquid surface rupture by a uniform electric field*. Physical Review, 1935. 48(6): pp. 562–568.

[13] Landau, L.D., Course of Theoretical Physics; v. 8., E.M. Lifshitz, Editor. 1960, Oxford: Reading, Mass: Pergamon Press; Addison-Wesley (*Electrodynamics of Continuous Media/by L.D. Landau and E.M. Lifshitz; translated from the Russian by J.B. Skyes and J.S. Bell.*).

[14] Lukas, D., A. Sarkar, and P. Pokorny, *Self-organization of jets in electrospinning from free liquid surface: ageneralized approach*. Journal of Applied Physics, 2008. 103(8) pp. 084309-1–084309-7.

[15] Lukáš, D., et al., *Physical principles of electrospinning (Electrospinning as a nano-scale technology of the twenty-first century)*. Textile Progress, 2009. 41(2): 59–140.

[16] Dao, A.T., *The Role of Rheological Properties of Polymer Solutions in Needleless Electrostatic Spinning*, in *PhD. thesis*. 2010, Technical University of Liberec: Liberec. p. 80.

[17] Yalcinkaya, F., et al., *Preparation of antibacterial nanofibre/nanoparticle covered composite yarns*. Journal of Nanomaterials, 2016. 2016(4): p. 7.

[18] Shenoy, S.L., et al., *Role of chain entanglements on fiber formation during electrospinning of polymer solutions: good solvent, non-specific polymer-polymer interaction limit*. Polymer, 2005. 46(10): pp. 3372–3384.

[19] McKinley, G.H. and T. Sridhar, *Filament-stretching rheometry*. Annual Review of Fluid Mechanics, 2002. 34: pp. 375–415.

[20] Barnes, H.A., *A Handbook of Elementary Rheology, University of Wales Institute of Non-Newtonian Fluid Mechanics*. 2000, University of Wales, Institute of Non-Newtonian Fluid Mechanics, Aberystwyth.

[21] Kim, K.W., et al., *The effect of molecular weight and the linear velocity of drum surface on the properties of electrospun poly(ethylene terephthalate) nonwovens*. Fibers and Polymers, 2004. 5(2): pp. 122–127.

[22] Buchko, C.J., et al., *Processing and microstructural characterization of porous biocompatible protein polymer thin films*. Polymer, 1999. 40(26): pp. 7397–7407.

[23] Mit-uppatham, C., M. Nithitanakul, and P. Supaphol, *Ultrafine electrospun polyamide-6 fibers: effect of solution conditions on morphology and average fiber diameter*. Macromolecular Chemistry and Physics, 2004. 205(17): pp. 2327–2338.

[24] Cengiz, F., T.A. Dao, and O. Jirsak, *Influence of solution properties on the roller electrospinning of poly(vinyl alcohol)*. Polymer Engineering & Science, 2010. 50(5): pp. 936–943.

[25] Fong, H., I. Chun, and D.H. Reneker, *Beaded nanofibers formed during electrospinning*. Polymer, 1999. 40(16): pp. 4585–4592.

[26] Filatov, Y., A. Budyka, and V. Kirichenko, *Electrospinning of Micro-and-Nanofibers: Fundementals and Applications in Seperation and Filtration Process*. 2007. Bgell House Inc. Danbury, Connecticut, United states.

[27] Kim, S.J., C.K. Lee, and S.I. Kim, *Effect of ionic salts on the processing of poly(2acrylamido-2-methyl-1-propane sulfortic acid) nanofibers*. Journal of Applied Polymer Science, 2005. 96(4): pp. 1388–1393.

[28] Heikkila, P. and A. Harlin, *Electrospinning of polyacrylonitrile (PAN) solution: effect of conductive additive and filler on the process*. Express Polymer Letters, 2009. 3(7): pp. 437–445.

[29] Yener, F. and O. Jirsak. *Improving Performance of Polyvinyl Butyral Electrospinning*. in *3rd International Conference on Nanocon*. TANGER LTD 2011. Brno, Czech Republic.

[30] Yener, F. and O. Jirsak, *Comparison between the Needle and Roller Electrospinning of Polyvinylbutyral*. Journal of Nanomaterials, 2012. 2012: p. 6.

[31] Yener, F. and O. Jirsak. *Development of New Methods for Study of Mechanism of Electrospinning*. in *Workshop*. 2012. Světlanka: Technical University of Liberec.

[32] Melcher, J.R. and G.I. Taylor, *Electrohydrodynamics—a review of role of interfacial shear stresses*. Annual Review of Fluid Mechanics, 1969. 1: pp. 111–146.

[33] Zeleny, J., *Instability of electrified liquid surfaces*. Physical Review, 1917. 10: pp. 1–6.

[34] Taylor, G.I., *Disintegration of water drops in an electric field*. Proceedings of the Royal Society A: Mathematical, Physical and Engineering Sciences, 1964. 280: pp. 383–397.

[35] Yarin, A.L., S. Koombhongse, and D.H. Reneker, *Taylor cone and jetting from liquid droplets in electrospinning of nanofibers*. Journal of Applied Physics, 2001. 90(9): pp. 4836–4846.

[36] Baumgart,P.K., *Electrostatic spinning of acrylic microfibers*. Journal of Colloid and Interface Science, 1971. 36(1): pp. 71–79.

[37] Deitzel, J.M., et al., *The effect of processing variables on the morphology of electrospun nanofibers and textiles*. Polymer, 2001. 42(1): pp. 261–272.

[38] Yuan, X.Y., et al., *Morphology of ultrafine polysulfone fibers prepared by electrospinning*. Polymer International, 2004. 53(11): pp. 1704–1710.

[39] Geng, X.Y., O.H. Kwon, and J.H. Jang, *Electrospinning of chitosan dissolved in concentrated acetic acid solution*. Biomaterials, 2005. 26(27): pp. 5427–5432.

[40] Zhao, Z.Z., et al., *Preparation and properties of electrospun poly(vinylidene fluoride) membranes*. Journal of Applied Polymer Science, 2005. 97(2): pp. 466–474.

[41] Zhang, C.X., et al., *Study on morphology of electrospun poly(vinyl alcohol) mats*. European Polymer Journal, 2005. 41(3): pp. 423–432.

[42] Yalcinkaya, F., B. Yalcinkaya, and O. Jirsak, *Analysis of the effects of rotating roller speed on a roller electrospinning system*. Textile Research Journal, 2016 doi: 10.1177/0040517516641362.

[43] Niu, H.T., X.G. Wang, and T. Lin, *Upward needleless electrospinning of nanofibers*. Journal of Engineered Fibers and Fabrics, 2012. 7: pp. 17–22.

[44] Reneker, D.H. and I. Chun, *Nanometre diameter fibres of polymer, produced by electrospinning*. Nanotechnology, 1996. 7(3): pp. 216–223.

[45] Shin, Y.M., et al., *Experimental characterization of electrospinning: the electrically forced jet and instabilities*. Polymer, 2001. 42(25): pp. 9955–9967.

[46] Theron, S.A., E. Zussman, and A.L. Yarin, *Experimental investigation of the governing parameters in the electrospinning of polymer solutions*. Polymer, 2004. 45(6): pp. 2017–2030.

[47] Fallahi, D., et al., *Effects of feed rate and solution conductivity on jet current and fiber diameter in electrospinning of polyacrylonitrile solutions*. E-Polymers, 2009. 9: pp. 1250–1257.

[48] Demir, M.M., et al., *Electrospinning of polyurethane fibers*. Polymer, 2002. 43(11): pp. 3303–3309.

[49] Bhattacharjee, P.K., et al., *On the measured current in electrospinning*. Journal of Applied Physics, 2010. 107(4): pp. 044306–1–044306–7.

[50] Kim, S.J., K.M. Shin, and S.I. Kim, *The effect of electric current on the processing of nanofibers formed from poly(2-acrylamido-2-methyl-1-propane sulfonic acid)*. Scripta Materialia, 2004. 51(1): pp. 31–35.

[51] Cengiz-Çallioğlu, F., O. Jirsak, and M. Dayik, *Electric current in polymer solution jet and spinnability in the needleless electrospinning process*. Fibers and Polymers, 2012. 13(10): pp. 1266–1271.

[52] Khumalo, M.B. and O. Jirsak, *Roller Electrospinning with Regards to Roller Movement*, in *MSc. Thesis, Nonwoven Dept*. 2012, Technical University of Liberec: Liberec. p. 60.

[53] Frenkel, A.L., et al., *Annular flows can keep unstable films from breakup-nonlinear saturation of capillary instability*. Journal of Colloid and Interface Science, 1987. 115(1): pp. 225–233.

[54] Campanella, O.H. and R.L. Cerro, *Viscous-flow on the outside of a horizontal rotating cylinder-the roll coating regime with a single fluid*. Chemical Engineering Science, 1984. 39(10): pp. 1443–1449.

[55] Jirsak, O., K. Frana, and F. Yener, *Electrospinning Studies: Jet Forming Force*, in *Strutex*, Faculty of Textile engineering. 2012. Liberec. p. 97.

[56] Deitzel, J.M., et al., *Key parameters influencing the onset and maintenance of the electrospinning jet*. Polymeric Nanofibers, 2006. 918: pp. 56–73.

[57] Deliu, R., et al., *Needleless electrospinning relaxation time of the aqueous solutions of poly (vinyl alcohol)*. Materiale Plastice, 2014. 51(1): pp. 62–66.

Chapter 5

Fabrication of Silk Fibroin Nanofibres by Needleless Electrospinning

Nongnut Sasithorn, Lenka Martinová,
Jana Horáková and
Rattanaphol Mongkholrattanasit

Additional information is available at the end of the chapter

http://dx.doi.org/10.5772/65835

Abstract

Silk fibroin nanofibres were fabricated using a needleless electrospinning technique. The procedure focused on a new method for the preparation of a spinning solution from silk fibroin. The role of the concentration of silk fibroin solution, applied voltage and spinning distance were investigated as a function of the morphology of the obtained fibres and the spinning performance of the electrospinning process. The biocompatibility of the obtained fibre sheets was evaluated using an *in vitro* testing method with MG-63 osteoblasts. The solvent system consisted of formic acid and calcium chloride that can dissolve silk fibroin at room temperature, and a rate of 0.25 g of calcium chloride per 1 g of silk fibroin was required to obtain a completely dissolved silk fibroin solution. The diameters of the silk electrospun fibres obtained from the formic acid–calcium chloride solvent system ranged from 100 to 2400 nm, depending on the spinning parameters. Furthermore, increasing the concentration of the silk fibroin solution and the applied voltage improved spinning ability and spinning performance in needleless electrospinning. In addition, *in vitro* tests with living cells showed that the obtained electrospun fibre sheets were highly biocompatible with MG-63 osteoblasts.

Keywords: silk fibroin, needleless electrospinning, formic acid, calcium chloride

1. Introduction

Silk is a fibrous protein that is produced by a variety of insects, including the silkworm, and silk fibres from silkworms have been used in textiles for almost 5000 years. The primary reasons for this enduring use have been the unique lustre, tactile properties, high mechanical strength, elasticity, durability, softness and dyeability of silks. Silks also display interesting thermal and electromagnetic responses, particularly in the UV range, for insect entrapment, and form crystalline phases related to processing [1]. In addition to its outstanding mechanical properties, silk is a candidate material for biomedical applications, as it has high biological compatibility and oxygen and water vapour permeability, in addition to being biodegradable and having minimal inflammatory reactions [2–5]. Silks have historically been employed in medicine as sutures during the past 100 years, and are still currently used in this manner, as well as in a variety of consumer product applications. Commercially, silkworm cocoons are mass-produced in a process termed 'sericulture' [1]. Although the silkworm spins its cocoon from a continuous filament of silk, the remainder of the silk cocoon is unsuitable for reeling and is known as silk waste. Silk waste also encompasses waste silk from all stages of production, from reeling through weaving. The composition of silk waste is similar to that of good silk. It has been roughly characterised by scientists, and has shown a high level of remaining nutrients, such as protein and lipid, which can be transformed into high-value products, for example, cosmetics, medical materials for human health, and food additives [6].

The silk fibroin protein is a structuring molecule that can be processed into numerous forms using a variety of techniques. Several different material morphologies can be formed from aqueous or organic solvent formulations of the natural fibre form of silk for utilisation in biomaterials for biomedical applications. Natural silk fibres dissolve only in a limited number of solvents because of the presence of a large amount of intra- and intermolecular hydrogen bonds in fibroin and its high crystallinity. Dissolution methods used for solubilising degummed silk fibroin fibres generally rely on strong chaotropic agents, including concentrated acids (hydrochloric acid, phosphoric acid and sulphuric acid) and aqueous salt solutions in high ionic strength (such as lithium thiocyanate, lithium bromide, calcium chloride, zinc chloride and magnesium chloride) to neutralise the hydrogen bonds stabilising the silk crystal structure. Consequently, the conditions of the dissolution process can influence the chemical composition and the molecular structure of the silk protein, affecting its biomaterial properties [7, 8].

The most ubiquitous method of producing regenerated silk fibroin solution is through the use of heavy salts, such as LiBr or $CaCl_2$. The degummed fibres can be dissolved using LiBr (9.3 M) solution or a ternary solvent system of $CaCl_2/C_2H_5OH/H_2O$ (1/2/8 mole ratio) to disrupt hydrogen bonding between the fibroin protein chains. The solution should then be allowed to thoroughly dissolve for up to 4 hours at 60°C to ensure complete dissolution. The heavy salts can then be removed from the silk solution through dialysis against deionised water over a period of 72 hours. Typically, the molecular weight cut-off for the dialysis membrane is 3500 Da, which is sufficiently permeable to allow the salts and water to travel freely, while retaining the fibroin light and heavy protein chains, respectively. The main disadvantage of a salt-containing aqueous solvent is the long preparation time, as aqueous solutions of fibroin must be dialysed for several days in order to remove the salts and recover the polymer as films, sponges or powder from the aqueous solution by dry forming. In some organic solvents (e.g. hexafluoroacetone and hexafluoroisopropanol), silk fibroin can be dissolved only after preliminary activation by dissolution in aqueous salt systems followed by recovery [7–9].

The electrospinning of silk solution is a favoured processing methodology for producing nanometre- to micron-scale fibres that result in a high degree of available surface area for use in creating scaffolds for tissue engineering and regenerative medicine purposes [9]. Electrospinning technology can be divided into two branches: needle electrospinning and needleless electrospinning. The needle electrospinning setup normally comprises a high-voltage power supply, and a syringe needle connected to a power supply and a collector. During the electrospinning process, a high electric voltage is applied to the polymer solution. When the electric force overcomes the surface tension of the polymer solution, the latter is ejected off the tip of the Taylor cone to form a polymer jet [10, 11]. As a needle can produce only one polymer jet, needle electrospinning systems have very low productivity, typically less than 0.3 g/h per needle, making them unsuitable for practical uses [10]. However, needleless electrospinning systems have recently been developed, in which, instead of the generation of a polymer jet from the tip of the needle, polymer jets form from the surface of free liquid by self-organisation [11–19]. For example, Jirsak et al. [14] invented a needleless electrospinning system using a roller as the fibre generator. The roller electrospinning device contains a rotating cylinder electrode, which is partially immersed in a polymer solution reservoir. When the roller slowly rotates, the polymer solution is loaded onto the upper roller surface. Upon application of a high voltage to the electrospinning system, a number of solution jets are simultaneously generated from the surface of the rotating spinning electrode, thereby improving fibre productivity [10].

In order to utilise silk waste, as well as to achieve large-scale production of electrospun fibre sheets, the regeneration of silk fibres is necessary, via a simple, but efficient, spinning process. A needleless electrospinning technique was chosen for fabrication of silk fibroin electrospun fibre sheets in the present study, as a result of its capacity to fabricate nanofibre layers on a mass industrial scale. However, needleless electrospinning is a new technique, and the majority of previous research on electrospinning of silk fibroin has focused on the needle electrospinning system. There is little information relating to the electrospinning of silk fibroin with a needleless system; therefore, the parameters of the spinning process have not been identified. The aim of the present research was to fabricate silk fibroin nanofibres with a needleless electrospinning method, and the experiment intensively concentrated on the effect of various parameters on the electrospinning process. The role of the concentration of silk fibroin solution, applied voltage and working distance were investigated as a function of the morphology of the obtained fibres and the spinning performance of the electrospinning process. In addition, a new method for preparation of the spinning solution, namely, dissolving silk fibroin in a mixture of formic acid and calcium chloride, was used for solution preparation instead of a ternary solvent system of $CaCl_2/C_2H_5OH/H_2O$ [20], which has been widely used to dissolve silk fibroin. Furthermore, characterisation of properties of the obtained electrospun fibre sheets and their interaction with living cells was also carried out.

2. Experiment

2.1. Materials

Waste cocoons of *Bombyx mori* Linn. Thai silkworm (Nang-Noi Srisakate 1) were supplied from Chan farm, Amphoe Mueang Chan, Si Sa Ket Province, Thailand, and waste silk cocoons, in the form of pierced cocoons, were obtained from the supplier. FBA-free ECE phosphate reference detergent (Union TSL Co., Ltd., Thailand) was used as a soaping agent in the degumming process. The chemicals used for the preparation of the spinning solutions were calcium chloride and 98% formic acid. All other chemicals used in this study were reagent grade.

2.2. Preparation of silk fibroin solutions

Silk fibroin (SF) was obtained using a degumming process. In brief, pierced cocoons were cleaned by first manually removing the contaminants and then washing twice with water. The cleaned raw cocoons were then dried in the sun. Raw silk cocoons were degummed twice with 1 M sodium carbonate (Merck Ltd.) and 0.5% of a soaping agent at 100°C for 30 minutes. Following degumming, the silk fibroin was washed in boiling water for 30 minutes to remove any remaining sericin from the surface of the fibre, and then washed again with warm water and dried at room temperature. Silk fibroin solutions were prepared by dissolving the degummed silk fibres in a mixture of formic acid and calcium chloride. The silk fibroin concentration varied from 6 to 14 wt%. All solutions were magnetically stirred at room temperature overnight.

2.3. Investigation of the effect of calcium chloride on the dissolution behaviour of silk fibroin in formic acid

In order to ascertain the appropriate amount of calcium chloride for dissolution of silk fibroin in formic acid, degummed silk fibres were directly dissolved in formic acid (98%) with various amounts of calcium chloride to prepare 12 wt% of silk fibroin solution. The weight ratio of silk fibroin to calcium chloride was 1:0.20, 1:0.25, 1:0.30, 1:0.35 and 1:0.40 (w/w), respectively. All solutions were magnetically stirred at room temperature and then electrospun into fibre sheets.

2.4. The needleless electrospinning process

A schematic representation of the equipment used in the spinning process is presented in **Figure 1**. The electrospinning device contained a rotating electrode, which was made of stainless steel wire of 125 mm in length and 0.02 mm in diameter (**Figure 2**), and a solution reservoir. The solution reservoir, which had a high voltage connected to the bottom of the solution bath, was filled with the silk fibroin solution. The process parameters are shown in **Table 1**. The electrospun fibres sheet was collected on the backing material moving along the collector electrode at a velocity of 10 mm/min.

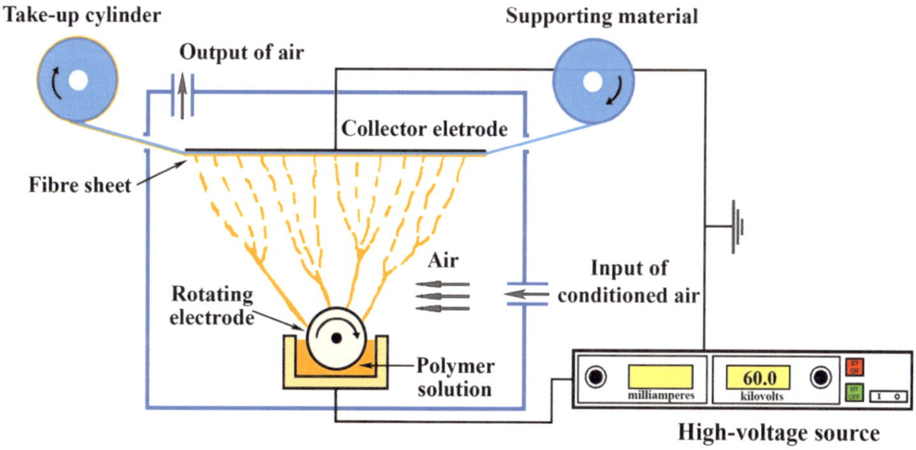

Figure 1. Schematics of the needleless electrospinning setup.

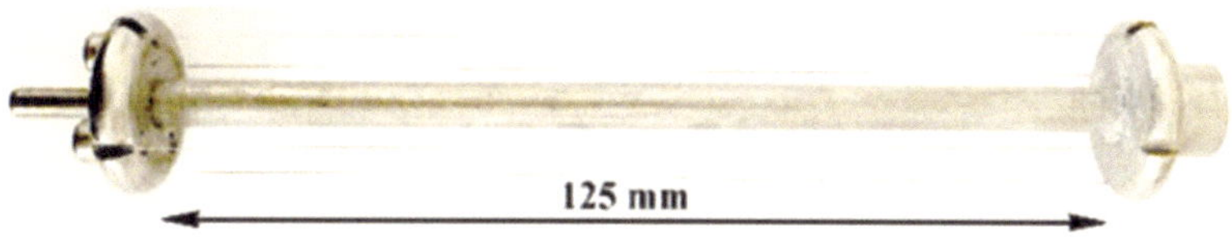

Figure 2. A spinning electrode.

Process parameters	Parameters level
Silk fibroin concentration (wt%)	6, 8, 10,12, 14
Applied voltage (kV)	30, 40, 50, 60
Distance between electrodes (mm)	100, 125, 150
Air humidity (%)	35–40
Temperature (°C)	20–25

Table 1. The spinning parameters of an electrospinning experiment.

2.5. Post-treatment of electrospun fibre sheets

Electrospun fibre sheets were treated with alcohol to achieve the solvent-induced crystallisation of the silk fibroin and to reduce the water solubility of the fibre sheets. The obtained fibre sheets were immersed in absolute ethanol for 30 minutes. After drying at room temperature, the treated fibre sheets were removed from the backing substrate and immersed in distilled water overnight, after which they were rinsed in distilled water to remove residual salts and then finally dried again.

2.6. Characterisation

2.6.1. Morphology analysis and fibre diameter

The morphological appearance of the electrospun fibres was observed with a scanning electron microscope (Vega 3, Tescan, Czech Republic) at an accelerated voltage of 20 kV. All the samples were sputter-coated (Q150R ES, Quorum Technologies Ltd., England) with gold at a thickness of 7 nm. The diameter of the nanofibres was measured by counting image pixels with NIS-Elements AR image software (LIM s.r.o., Czech Republic). The average fibre diameter and distribution were determined from 200 random fibres obtained under each spinning condition.

2.6.2. Spinning performance of the electrospinning process

Spinning performance is one of the most important parameters of the needleless electrospinning method. It was calculated from the mass per unit area and width of the electrospun fibre sheets and the velocity of the backing material, using the equation 1 [16]:

$$P = \frac{G \times W \times V_f}{L_r} \tag{1}$$

where P is a spinning performance (g/min/m), G is a mass per unit area of electrospun fibres sheet in grams per square metre (g/m^2), W is width of fibre layers in metres (m), V_f is take-up cylinder speed in metres per minute (m/min) and L_r is length of spinning electrodes in metres (m).

2.7. *In vitro* tests of silk electrospun fibre sheets

2.7.1. Preparation of scaffolds and cell seeding

The fibre sheets were cut into small disks with a diameter of 6 mm to fit each well of a 96-well plate. The specimens were sterilised by immersion in a 70% aqueous ethanol solution for 30 minutes, followed by double washing in phosphate-buffered saline (PBS, Lonza). MG-63 osetoblasts were cultivated in Eagle's Medium (Lonza) supplemented by 10% foetal bovine serum and 1% antibiotics. The cells were placed in a humidified incubator at an atmosphere of 5% CO_2 at 37°C. When they became confluent, the cells were suspended using trypsin-EDTA, centrifuged and re-suspended in fresh complete medium. The number of cells was determined by a Luna™ cell counter. Osteoblasts were seeded on the scaffolds and placed in a 96-well plate at a density of 5×10^3 per well plate. The medium was changed three times per week during the experiment.

2.7.2. Cell adhesion and proliferation analysis

The viability of the cells seeded on the scaffolds was analysed via MTT test after 1, 3, 7 and 14 days of culture. In brief, after cell culture, 50 μl MTT solution (Sigma Aldrich) was added to

http://dx.doi.org/10.5772/65835

150 μl of Dulbecco's Modified Eagle's Medium (DMEM) and the specimens were incubated at 37°C for 4 hours. When MTT (3-(4,5-dimethylthiazol-2-yl)-2,5-diphenyl-2*H*-tetrazolium bromide) was reduced to purple formazan by mitochondrial dehydrogenase in the cells, this indicated normal metabolism. The formed violet crystals of formazan were then solubilised with acidic isopropanol, and the optical density of the suspension was measured using an Absorbance Reader ELx808 (BioTek). For each testing day, four samples of each material were incubated with MTT solution and average absorbance was calculated as the difference between absorbance measured at 570 nm and by reference wavelength of 690 nm.

2.7.3. Fluorescence microscopy analysis

After the culture period, the samples were washed twice in PBS and fixed in frozen methanol for 30 minutes followed by double washing with PBS and staining with propidium iodide (PI dilution 2 g/l PBS, Sigma Aldrich) for 10 minutes in the dark. The stained cells were observed by inverted fluorescence microscope (Nikon).

2.7.4. Cell morphology

The cell morphology was analysed using a scanning electron microscope (SEM). After days of culture the samples were washed twice in PBS and fixed in 2.5% glutaraldehyde in PBS for 30 minutes at 4°C. The samples were then dehydrated by gradients of ethanol (60, 70, 80, 90, 96 and 100%, respectively). After water removal, the scaffolds were transferred to the SEM holder, sputter-coated with gold and observed under the SEM.

3. Results and discussion

3.1. The effect of calcium chloride on the dissolution behaviour of silk fibroin

3.1.1. The effect of calcium chloride on solubility of silk fibroin in formic acid

Figure 3 shows the difference in dissolution of silk fibroin in formic acid with varying amounts of calcium chloride. It was observed that silk fibroin was insoluble in formic acid, but was soluble in a mixture of formic acid and calcium chloride. This shows that calcium chloride can improve the solubility of silk fibroin in formic acid. It is suggested that calcium chloride has a chaotropic property that disrupts stabilising intra-molecular forces, such as hydrogen bonds, van der Waals forces and hydrophobic interactions, in protein structures by shielding charges. Hydrogen bonds are stronger in non-polar media. Calcium chloride that increases the chemical polarity of the solvent can also destabilise hydrogen bonding in the silk fibroin structure and ion-dipole interactions between the salts and hydrogen bonding species, which are more favourable than normal hydrogen bonds. This increases the solubility of hydrophobic proteins in the solvent [7, 21]. The results showed that silk fibroin completely dissolved in formic acid when the weight ratio of 1:0.25 (w/w) of silk fibroin to calcium chloride was used.

Figure 3. The dissolution of silk fibroin in formic acid with varying amounts of calcium chloride: (a) without $CaCl_2$; (b) 1:0.15; (c) 1:0.20; (d) 1:0.25 [SF:$CaCl_2$ (w/w)].

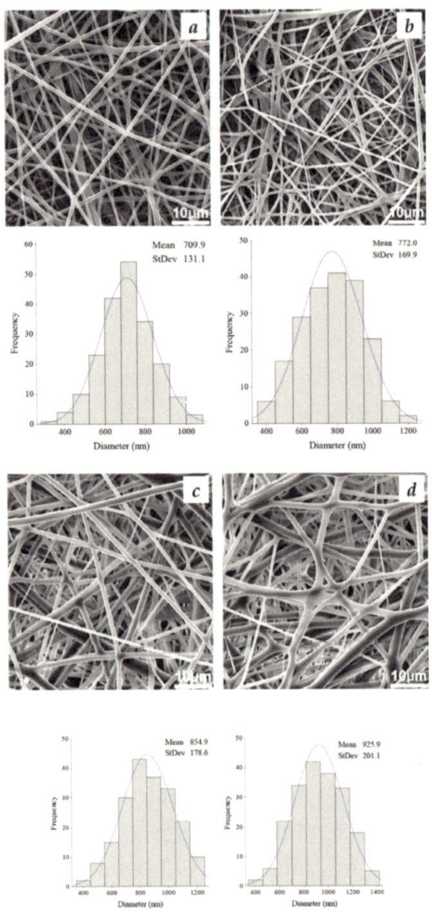

Figure 4. SEM micrographs and diameter distribution of electrospun fibres prepared from silk fibroin 12 wt% with varying amounts of $CaCl_2$: (a) 1:0.25; (b) 1:0.30; (c) 1:0.35; (d) 1:0.40; (SEM magnification 5×).

3.1.2. The effect of calcium chloride on the morphology of silk fibroin electrospun fibres

The SEM micrographs and diameter distribution of the silk fibroin electrospun fibres prepared from silk solutions with varying amounts of calcium chloride are shown in **Figure 4**. The results showed that the average fibre diameter increased with an increase in the amount of calcium chloride. Moreover, increasing the weight ratios of silk fibroin to calcium chloride affected the morphology of the obtained fibres; the fibres changed slightly from circular cross section to flat. It is assumed that the addition of a large quantity of calcium chloride can result in a change in the evaporation of solvent. Calcium chloride has the capacity to attract moisture from the air and surroundings, so the capacity of the silk fibroin solution to absorb water was increased when the amount of salt was increased. As a result, the solution absorbed more ambient water during electrospinning. The absorption of water does not allow completion of the drying process during the time of flight of the solution jet, and this phenomenon could lead to a slowing in evaporation of the solvent, which may result in an increase in fibre diameter and produce congealed mats instead of unwoven fibres [22, 23]. Therefore, the weight ratios of 1:0.25 and 1:0.30 (w/w) of silk fibre to calcium chloride appear to be suitable for the preparation of silk fibroin solution, which is used for an electrospinning process. The weight ratio of 1:0.25 (w/w) was eventually preferred and used throughout the remaining experiments.

3.2. The effect of parameters on needleless electrospinning of silk fibroin

3.2.1. The effect of silk fibroin concentration

The effect of concentration of the solution on fibre morphology with the applied voltage of 60 kV when the silk fibroin concentration increased from 6 to 14 wt% was considered. The effect of silk concentration on the morphological appearance and diameter of the electrospun fibres was investigated by SEM, as shown in **Figure 5**. It was found that an increase in the concentration of silk fibroin solution produced a significant effect on the average fibre diameter and the uniform diameter distribution of the obtained electrospun fibre. The results showed that the fibre diameter and non-uniform fibre diameter distribution of the obtained electrospun fibres increased with an increase in the silk fibroin concentration, showing the important role of the concentration of the silk fibroin solution in fibre formation during the needleless electrospinning process. When the concentration was increased from 6 to 14 wt%, the average fibre diameter increased from 191 to 1500 nm, respectively. The concentration of the polymer solution reflects the number of entanglements of polymer chains in the solution, which, in turn, affected the viscosity of the solution. An increase in the concentration of the silk solution resulted in greater polymer chain entanglement. Thus, the viscosity of the solution also increased. At higher concentrations, the diameter of the fibre was greater. In addition, the interaction between the solution and the charges on the jet determined the distribution of the fibre diameters that were obtained. This was probably due to the number of jets that formed during electrospinning. Multiple jets may have formed from the main electrospinning jet, which was sufficiently stable to yield fibres of a smaller diameter at certain concentrations, thereby generating fibres with various diameters [24, 25].

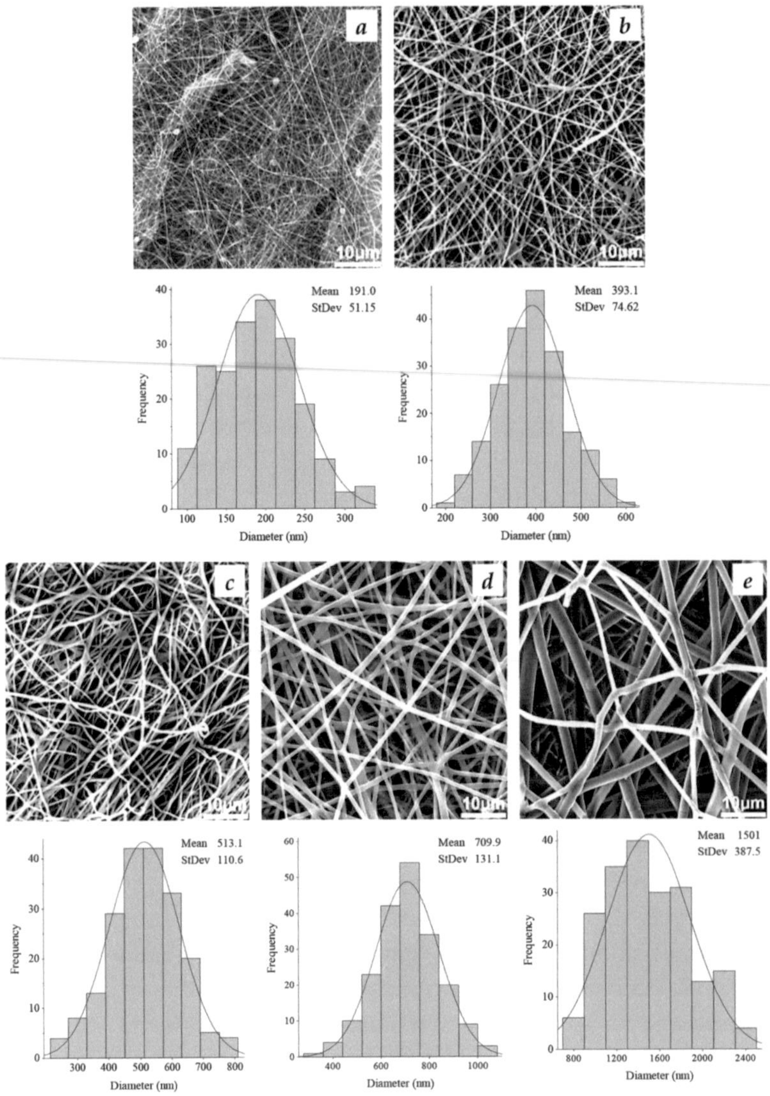

Figure 5. SEM micrographs and diameter distribution of electrospun fibres produced by needleless electrospinning with silk fibroin solution at various concentrations : (a) 6 wt%; (b) 8 wt%; (c) 10 wt%; (d) 12 wt%; (e) 14 wt% (magnification 5×).

In this study, the concentration of the silk solution played an important role in the spinnability of the needleless system. At low concentrations of spinning solution, non-fibrous formations were produced instead of nanofibres with beads. It is possible that Taylor cones are created in needleless electrospinning by picking up the spinning solution covering the surrounding spinning electrode [5, 14]. In low-concentration spinning solutions, the viscosity of the solution is also generally low. Such solutions cannot be loaded on the surface of the spinning electrode because of their lack of viscosity. When Taylor cones do not form on the surface of an electrode, the electrospinning process results in non-fibrous formations [24]. It was observed that no fibres were formed when a silk fibroin concentration of less than 6 wt% was used for this spinning condition. Although a silk fibroin solution with a 6 wt% concentration can spin into nanofibres with the needleless system, the spinning solution still has a low viscosity. Therefore, some droplets were observed on the obtained fibre sheet. A further increase in the concentration of silk fibroin up to 8 wt% resulted in continuous nanofibres and the droplets disappeared. This was due to the fact that there were sufficient molecular chain entanglements in the polymer solution to prevent the breakup of the electrically driven jet and to allow the electrostatic stresses to further elongate the jet to form fibres. Therefore, the silk fibroin solution required a concentration of at least 8 wt% to produce continuous silk fibres with a nanometre diameter under the experimental conditions of the present study.

In addition to affecting the fibre morphology and spinnability, the concentration of the spinning solution also influenced the spinning performance (see **Figure 6**). Under the same processing parameters, the spinning performance increased constantly from 0.22 to 2.05 g/min/m when the silk fibroin concentration increased from 6 to 14 wt%. The reason for this was the high solution viscosity, which facilitated jet/filament formation.

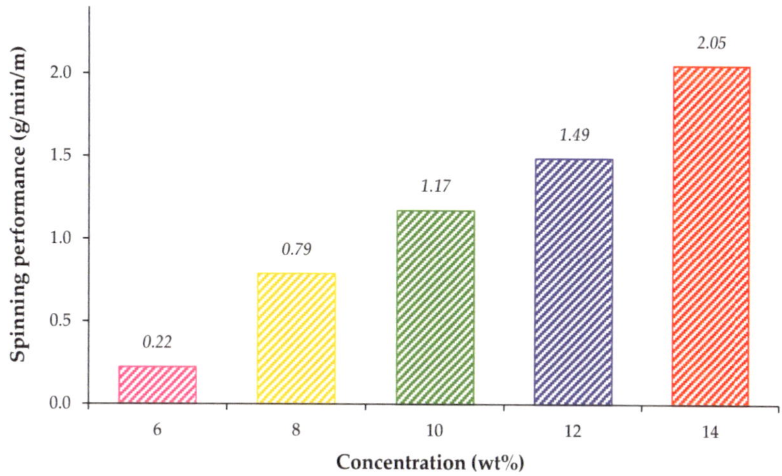

Figure 6. The effect of silk fibroin concentration on the spinning performance of the process.

3.2.2. *The effect of applied voltage*

The applied voltage is a very important parameter with regard to the formation of jets in electrospinning systems because a high voltage is used to create an electrically charged jet in a polymer solution [15, 18]. In order to evaluate the effect of the applied voltage on the electrospinning process, a spinning solution with a concentration of 12 wt% was electrospun at a voltage between 30 and 60 kV. SEM micrographs of the obtained fibre and their diameter distributions at the different voltages are shown in **Figure 7**. In the present study, the critical voltage required to initiate silk nanofibres in the needleless electrospinning system was higher than in the needle system [24]. When the silk fibroin solution was charged with an electric voltage higher than 26 kV, a number of jets were generated from the surface of the spinning electrode. The results showed that increasing the applied voltage produced an effect on the fibre diameter, but the concentration of silk fibroin solution appeared to have a greater effect on the fibre diameter than did the applied voltage. With an applied voltage increase from 30 to 60 kV, the average fibre diameter decreased from 940 to 710 nm, respectively.

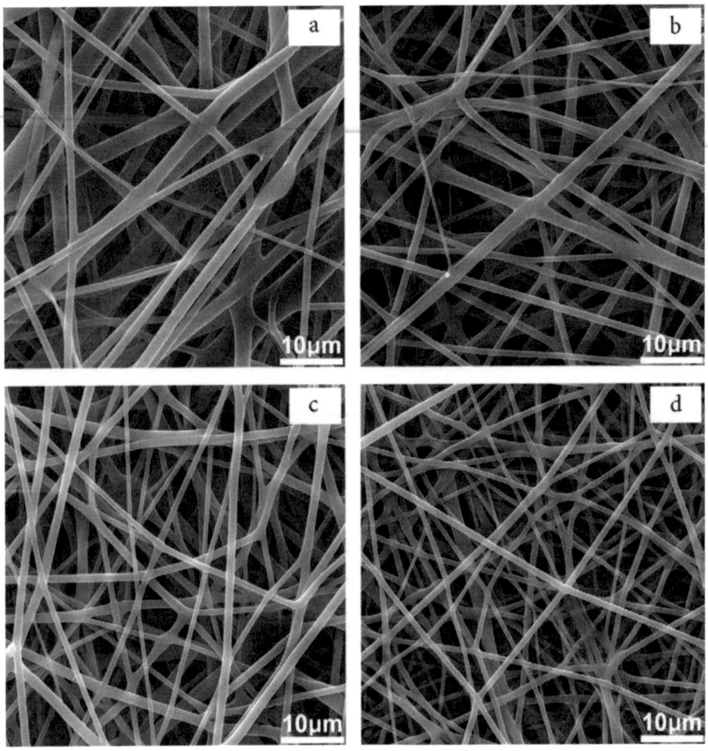

Figure 7. SEM micrographs and diameter distribution of electrospun fibres prepared by needleless electrospinning from silk fibroin 12 wt% at various applied voltages: (a) 30 kV; (b) 40 kV; (c) 50 kV; (d) 60 kV (SEM magnification 5×).

However, the spinning performance of the electrospinning process was influenced by the applied voltage and polymer concentration; it changed from 0.25 to 1.49 g/min/m, when the voltage was increased from 30 to 60 kV (**Figure 8**). As the electric field was the main driving force initiating the formation of Taylor cones and jets from the surface of the solution, increasing the electric voltage increased the electrostatic force on the polymer jet, which favoured further elongation of the jet and the formation of smaller fibres.

In contrast, the electric field also functioned to overcome the frictional forces that acted within the moving polymer solution and to accelerate filament movement towards the collector electrode. It is easier to generate solution jets at higher applied voltage in a polymer solution charged by a stronger electric field, because a larger amount of solution is removed from the surface of the solution, thereby improving the spinning performance of the process [12, 15, 17].

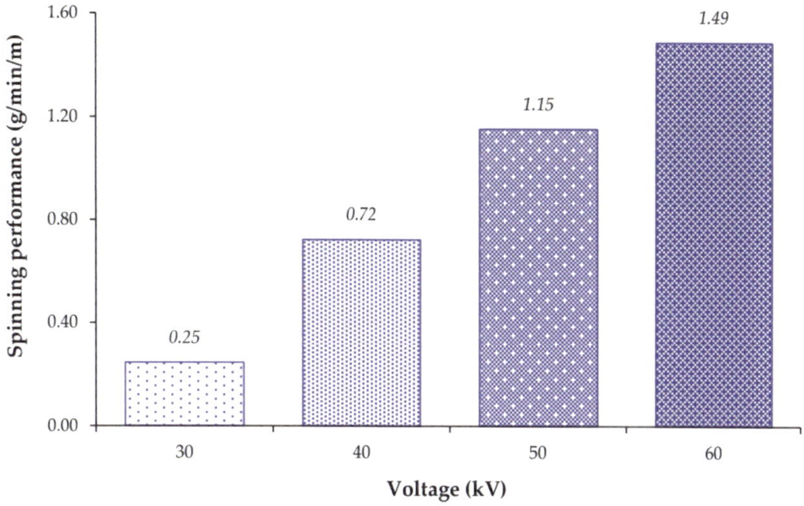

Figure 8. The effect of applied voltage on the spinning performance of the process.

3.2.3. The effect of distance between electrodes

In order to study the effect of spinning distance on the morphology of the obtained electrospun fibres and the spinning performance of the process, spinning solutions with a concentration of 12 wt% were electrospun at a high voltage of 60 kV. Electrospinning was carried out at a distance of 100, 125 and 150 mm. SEM micrographs of the resulting fibres and their distributions at the different spinning distances are shown in **Figure 9**. It was shown that an increase in spinning distance had a less significant effect on the average fibre diameter, but produced an effect on the spinning performance of the electrospinning process. With an increase in the distance from 100 to 150 mm, the average fibre diameter decreased from 710 to 647 nm, and the spinning performance of the electrospinning process changed from 1.49 to 0.70 g/min/m when the distance was increased (**Figure 10**). The distance between the spinneret and the collector is a key factor in determining the morphology of fibres and the spinning performances that are produced. It is suggested that increasing the distance has the same effect as decreasing the applied voltage, and that this will lead to a decrease in the field strength. As the electric field is the main driving force in initiation of the formation of jets from the surface of the solution, decreasing the electric voltage will decrease the electrostatic force on the polymer jet, resulting in a decrease of the spinning performance of the process. In other circumstances, increasing the distance results in a decrease in the average fibre diameter. As mentioned previously, jet elongation and thinning only occurs while the jet is in flight and still a fluid. This elongation occurs as a result of charge repulsion between ions in the solution combined with a net pull towards the collector. While in flight, the polymer solution solidifies as the solvent evaporates from the surface, forming polymer fibres. Therefore, increasing the spinning distance will increase the time for thinning to occur, and, provided the polymer is not yet solid, the fibre diameter will be reduced [22].

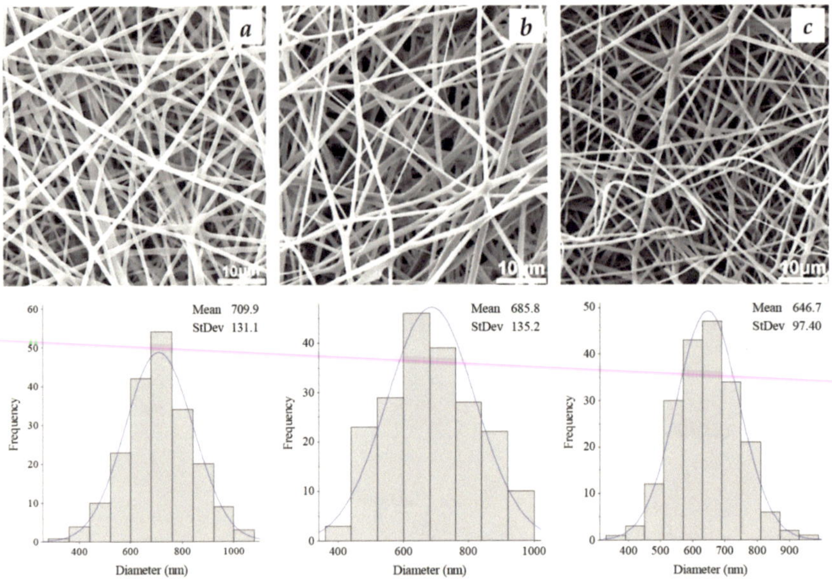

Figure 9. SEM micrographs and diameter distribution of electrospun fibres prepared by needleless electrospinning from silk fibroin 12 wt% at different spinning distance: (a) 100 mm; (b) 125 mm; (c) 150 mm (SEM magnification 5×).

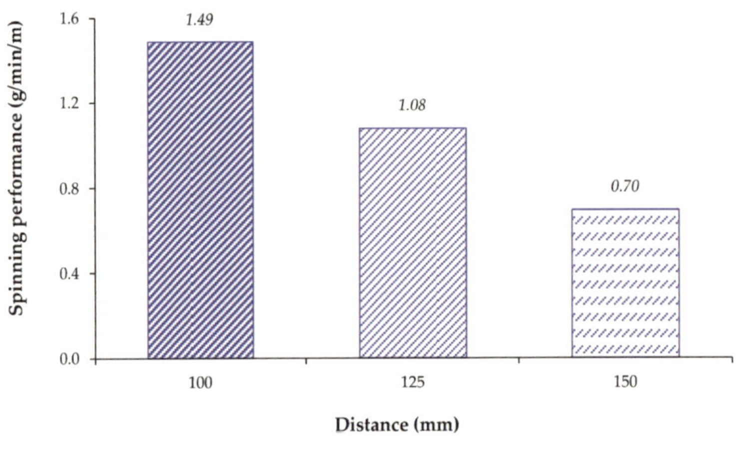

Figure 10. The effects of spinning distance on the spinning performance of the process.

3.3. *In vitro* cell interaction with MG-63 osteoblasts

The viability of cells seeded on the fibre sheets was measured by the MTT test during culture. The results are shown in **Figure 11**, and it can be seen that the tested materials supported osteoblast proliferation during 2 weeks of culture. The proliferation rate of MG-63 osteoblasts increased after 7 days of culture, whereas the absorbance increased. The highest proliferation rate was found between days 7 and 14.

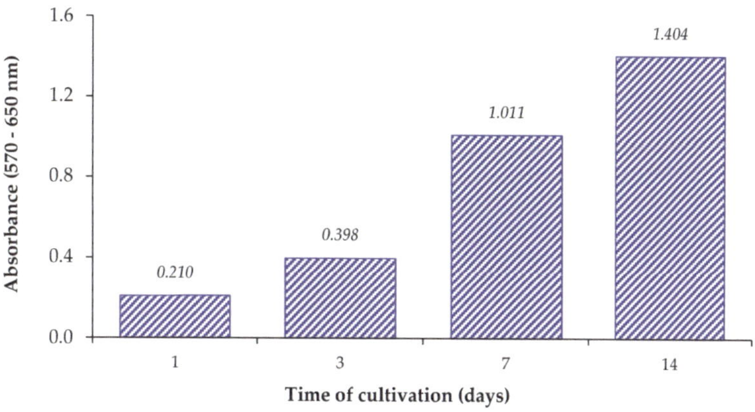

Figure 11. Cell viability measured by MTT test after culture with MG-63 osteoblasts.

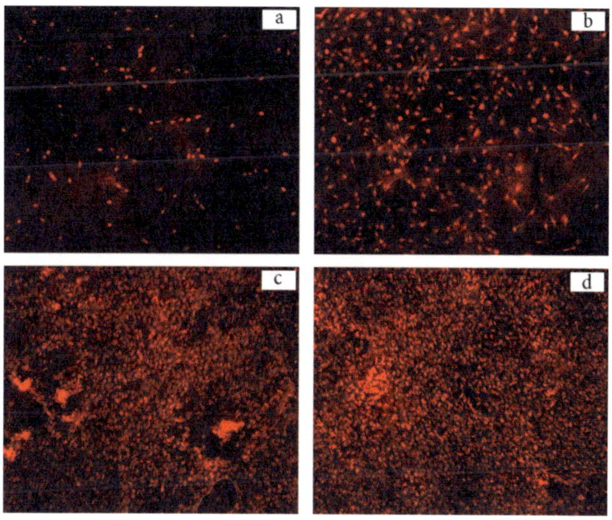

Figure 12. Fluorescence microscopy of MG-63 osteoblasts stained with propidium iodide during cell culture: (a) 1 day; (b) 3 days; (c) 7 days; (d) 14 days (magnification 100×).

As shown in **Figure 12**, the cells adhered and proliferated well on the surface of the fibre sheets during the time of culture. Fluorescence microscopy pictures confirmed the MTT test results that were obtained in the previous experiment. This showed a very high biocompatibility between silk fibroin and MG-63 osteoblasts.

The SEM micrographs from the scanning electron microscopy were in accordance with the fluorescence microscopy results, as well as the MTT test. As shown in **Figure 13**, the tested materials showed good adhesion when evaluated 1 day after seeding of osteoblasts onto the fibre sheets. During the 2 weeks of the experiment, the cells proliferated through the surface of the fibre sheets and covered most of their surface, and single cells can be distinguished. The results were confirmed by viability measurement with the MTT test and fluorescence microscopy, as well as scanning electron microscopy.

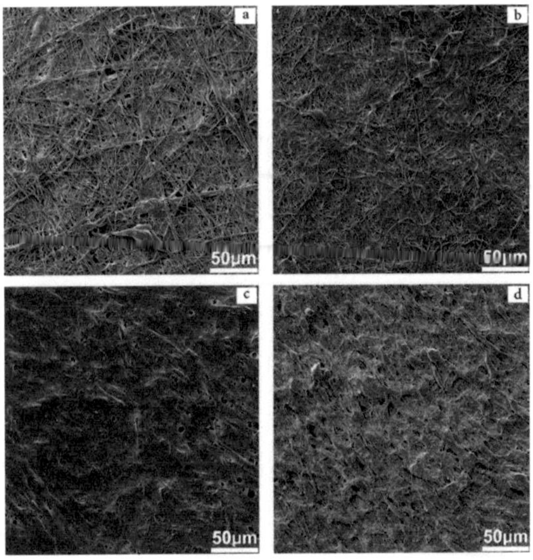

Figure 13. SEM micrographs of the fibre sheets after cells were cultured with MG-63 osteoblasts: (a) 1 day; (b) 3 days; (c) 7 days; (d) 14 days (SEM magnification 1×).

4. Conclusion

Silk fibroin nanofibres were electrospun with a needleless electrospinning method, focusing on the effect of the processing parameters on the morphology of the obtained fibres and the spinning performance of the electrospinning process. The use of formic acid-calcium chloride as the solvent for silk fibroin dissolution has the advantage of being simple in an operation. A solvent system consisting of formic acid and calcium chloride directly dissolved silk fibroin at room temperature. The weight ratio of 1:0.25 (w/w) of silk fibres to calcium chloride appears to be suitable for the dissolution of silk fibroin in formic acid. Thus, dissolution of silk fibroin in formic acid and calcium chloride could potentially be employed in the preparation of spinning solution for a large-scale production of silk nanofibres with a needleless electrospinning method.

In a needleless electrospinning process of silk fibroin in this solvent system, the concentration of the silk solution played an important role in the spinnability. Concentrations of silk fibroin ranging from 8 to 12 wt% appear to be suitable for the preparation of silk fibroin nanofibres with needleless electrospinning. Furthermore, increasing the concentration of the silk fibroin solution improved the spinning ability and the spinning performance of the electrospinning process. An increase in the applied voltage also enhanced the spinning performance of the process; however, an increase in the applied voltage had little effect on the reduction of the diameter of silk fibroin electrospun fibres. In contrast, the variation of spinning distance in the spinning process affected the spinning performance. The spinning performance of the process was decreased when the spinning distance was increased. Although the electrospinning technique and dissolution method are different from those used in previous studies, the results from *in vitro* tests with MG-63 osteoblasts showed very high biocompatibility with regard to silk fibroin electrospun fibre sheets. The fibre sheets were capable of promoting adhesion, spreading and proliferation. It can be assumed that the electrospun fibre sheet is a promising material for biomedical applications, such as bone-tissue engineering.

Acknowledgements

This work was supported by the Technical University of Liberec, Faculty of Textile Engineering, Czech Republic, and the Ministry of Education, Youth and Sports as part of Project LO1201 targeted support from the programme 'Národní program udržitelnosti I'. The authors also thank Rajamangala University of Technology Phra Nakhon, Thailand, for providing the first author with a scholarship.

Author details

Nongnut Sasithorn[1,2*], Lenka Martinová[3], Jana Horáková[2] and Rattanaphol Mongkholrattanasit[1]

*Address all correspondence to: nongnut.s@rmutp.ac.th

1 Textile Chemical Technology Department, Faculty of Industrial Textiles and Fashion Design, Rajamangala University of Technology Phra Nakhon, Bangkok, Thailand

2 Department of Nonwovens and Nanofibrous Materials, Faculty of Textile Engineering, Technical University of Liberec, Liberec, Czech Republic

3 Institute for Nanomaterials, Advanced Technology and Innovation, Technical University of Liberec, Liberec, Czech Republic

References

[1] Matsumoto A., Kim H.J., Tsai I.Y., Wang X., Cebe P., Kaplan D.L.. Silk. In: Lewin M., editor. Handbook of Fiber Chemistry. 3rd ed. Boca Raton: CRC Press Taylor & Francis Group; 2007. p. 383–404.

[2] Vepari C., Kaplan D.L. Silk as a Biomaterial. Progress in Polymer Science. 2007;32(8–9): 991–1007. DOI: 10.1016/j.progpolymsci.2007.05.013

[3] Min B.M., Lee G., Kim S.H., Nam Y.S., Lee T.S., Park W.H. Electrospinning of Silk Fibroin Nanofibers and Its Effect on the Adhesion and Spreading of Normal Human Keratinocytes and Fibroblasts *In Vitro*. Biomaterials. 2004;25(7–8):1289–1297. DOI: 10.1016/j.biomaterials.2003.08.045

[4] Amiralian N., Nouri M., Kish M.H. An Experimental Study on Electrospinning of Silk Fibroin [Internet]. Available from: http://www.docstoc.com/docs/26292086 [Accessed: 2011-05-02].

[5] Min B.M., Jeong L., Lee K.Y., Park W.H. Regenerated Silk Fibroin Nanofibers: Water Vapor-Induced Structural Changes and Their Effects on the Behavior of Normal Human Cells. Macromolecular Bioscience. 2006;6(4):285–292. DOI: 10.3201/eid1206.AD1206

[6] Prommuaka C., De-Eknamkulb W., Shotipruka A. Extraction of Flavonoids and Carotenoids from Thai Silk Waste and Antioxidant Activity of Extracts. Separation and Purification Technology. 2008;62(2):444–448. DOI: 10.1016/j.seppur.2008.02.020

[7] Sashina E.S., Bochek A.M., Novoselov N.P., Kirichenko D.A. Structure and Solubility of Natural Silk Fibroin. Russian Journal of Applied Chemistry. 2006;79(6):869–876. DOI: 10.1134/S1070427206060012

[8] Abdel-Naby W., Lawrence B.D. Processing of Silk Biomaterials. In: Basu A., editor. Advances in Silk Science and Technology. Cambridge: Woodhead Publishing; 2015. pp. 171–183. DOI: 10.1016/B978-1-78242-311-9.00009-4

[9] Lawrence B.D. Processing of *Bombyx mori* Silk for Biomedical Applications. In: Kundu S.C., editor. Silk Biomaterials for Tissue Engineering and Regenerative Medicine. Cambridge : Woodhead Publishing; 2014. pp. 78–98. DOI: 10.1533/9780857097064.1.78

[10] Niu H., Wang X., Lin T. Needleless Electrospinning: Developments and Performances. In: Lin T., editor. Nanofibers-Production, Properties and Functional Applications. Rijeka : InTech; 2011. pp. 17–36. DOI: 10.5772/24999

[11] Lukáš D., Sarkar A., Martinová L., Vodsedálková K., Lubasová D., Chaloupek J., et al. Physical Principles of Electrospinning (Electrospinning as a Nano-Scale Technology of The Twenty-First Century). Textile Progress. 2009;41(2):59–140. DOI: 10.1080/00405160902904641

[12] Niu H., Lin T., Wang X. Needleless Electrospinning I. A Comparison of Cylinder and Disk Nozzles. Journal of Applied Polymer Science. 2009;114(6):3524–3530. DOI: 10.1002/app.30891

[13] Yarin A.L., Zussman E. Upward Needleless Electrospinning of Multiple Nanofibers. Polymer. 2004;45(9):2977–2980. DOI: 10.1016/j.polymer.2004.02.066

[14] Jirsak O., et al. A Method of Nanofibers Production from a Polymer Solution Using Electrostatic Spinning and a Device for Carrying Out the Method, in EP1673493. Czech Republic; 2005.

[15] Wang X., Niu H., Lin T., Wang X. Needleless Electrospinning of Nanofibers with a Conical Wire Coil. Polymer Engineering and Science. 2009;49(8):1582–1586. DOI: 10.1002/pen.21377

[16] Yener F., Jirsak O. Comparison between the Needle- and Roller Electrospinning of Polyvinylbutyral. Journal of Nanomaterials. 2012;2012:839317. DOI: 10.1155/2012/839317

[17] Huang C., Niu H., Wu J., Ke Q., Mo X., Lin T. Needleless Electrospinning of Polystyrene Fibers with an Oriented Surface Line Texture. Journal of Nanomaterials. 2012;2012:473872. DOI: 10.1155/2012/473872

[18] Wang X., Niu H., Wang X., Lin T. Needleless Electrospinning of Uniform Nanofibers Using Spiral Coil Spinnerets. Journal of Nanomaterials. 2012;2012:785920. DOI: 10.1155/2012/785920.

[19] Cengiz F., Jirsak O. The Effect of Salt on the Roller Electrospinning of Polyurethane Nanofibers. Fibers and Polymers. 2009;10(2):177–184. DOI: 10.1007/s12221-009-0177-7

[20] Ajisawa A. Dissolution of Silk Fibroin with Calcium Chloride/Ethanol Aqueous Solution. Journal of Sericultural Science of Japan. 1998;67(2):91–94. DOI: 10.11416/kontyushigen1930.67.91

[21] Sasithorn N., Martinová L. Effect of Calcium Chloride on Electrospinning of Silk Fibroin Nanofibres. RMUTP Research Journal: Special Issue. 2014; 62–69.

[22] Robb B., Lennox B. The Electrospinning Process, Conditions and Control. In: Bosworth L.A., Downes S., editors. Electrospinning for Tissue Regeneration. Cambridge : Woodhead Publishing; 2011. pp. 51–66.

[23] De Vrieze S., Van Camp T., Nelvig A., Hagström B., Westbroek P., De Clerck K. The Effect of Temperature and Humidity on Electrospinning. Journal of Materials Science. 2009;44(5):1357–1362. DOI: 10.1007/s10853-008-3010-6

[24] Sasithorn N., Martinová L. Fabrication of Silk Nanofibres with Needle and Roller Electrospinning Methods. Journal of Nanomaterials. 2014;2014:947315. DOI: 10.1155/2014/947315

[25] Ramakrishna S., Fujihara K., Teo W-E., Lim T-C., Ma Z. An Introduction to Electrospinning and Nanofibers. Singapore: World Scientific Publishing Co. Pte. Ltd.; 2005. 396 p.

Section 3

Application of Electrospinnig in Biomedical Field

Chapter 6

Electrospinning in Tissue Engineering

Yawen Li and Therese Bou-Akl

Additional information is available at the end of the chapter

http://dx.doi.org/10.5772/65836

Abstract

Electrospinning employs a strong electric field to draw charged polymer fluids or melts into fibers with diameter in the range from tens of nanometers to microns. The relatively simple experimental setup, a wide range of suitable materials, and the possibility of incorporating bioactive molecules into the fibers make electrospinning a versatile process in creating scaffolds for tissue engineering applications. This chapter reviews the electrospinning process and discusses how solution and processing parameters affect the electrospun fiber structure and function. A brief overview of various surface modification methods used for enhancing the cell adhesion, proliferation, and differentiation on the fibrous scaffolds is provided. Commonly used methods include physical entrapment, chemical treatment, and coelectrospinning. The application of electrospun fibrous scaffolds in tissue engineering is reviewed, focusing on recent progress in the regeneration of skin, vasculature, bone, ligaments, and tendons.

Keywords: electrospinning process, electrospun fibers, scaffolds, surface modification, tissue engineering

1. Introduction

Electrospinning is closely related to the more established technology of electrospraying, a process in which electrostatic forces are used to control the formation of fluid droplets. The earliest description of electrospraying can be traced back to the early seventeenth century, when William Gilbert observed that a drop of water deformed in a cone in the presence of charged amber [1]. John Zeleny's work on the effect of an electric field on a liquid meniscus in the early twentieth century was considered as the beginning to the development of electrospraying and electrospinning technologies [2, 3]. In the mid-1960s, Sir Geoffrey Ingram Taylor published a series of articles and established the theoretical framework to understand the behavior of electrified fluids [4, 5]. The theory has been further developed and refined by more recent literature [6–9] that helps to better understand and guide the electrospraying/electrospinning process.

Electrospraying has been widely applied to develop commercial technologies such as ionization source for mass spectrometry, liquid metal ion source for ion implantation, focused ion beam instruments, and electrostatic precipitation of nanoparticles. Interest in electrospinning grew more slowly until mid-1990s, when work led by Reneker and coworker demonstrated the production of continuous nanofibers using electrospinning and predicted their potential applications in filtration, biology, energy conversion, and agriculture [10]. The past decade has seen exponential growth in electrospinning-related literature, totaling over 3000 articles and 1000 issued patents from 2001 to 2015 [11].

Tissue engineering is an emerging multidisciplinary field that integrates engineering principles with biology and medicine with the goal of restoring or enhancing tissue or organ functions [12]. One key component of a tissue-engineered construct is a porous biodegradable scaffold to provide structural support for the cells. The relatively simple and inexpensive setup makes electrospinning a versatile process with high production capability to form submicron sized nonwoven fibrous scaffolds. The possibility of incorporating one or multiple bioactive factors is another advantage of the process.

This chapter will review the electrospinning process, and discuss how solution and process parameters affect the morphology of different types of polymer fibers. The inherent hydrophobicity of many synthetic polymers leads to suboptimal cell adhesion and functions on electrospun fibrous scaffolds. We will discuss different surface modification methods to enhance cell adhesion, proliferation, and differentiation. Finally, we will provide some case studies to illustrate the application of electrospun fibers as scaffolds to regenerate a variety of tissues.

2. Electrospinning process control

2.1. Process overview

Electrospinning uses electrostatic forces to produce fibers from polymer solutions. One attractive feature of this process is the relatively simple and inexpensive experimental setup. A typical electrospinning apparatus (as shown in **Figure 1**) consists of three components: a spinneret (usually a metal hollow needle), a high voltage source, and a collector (grounded or negatively biased). A syringe pump is commonly used to drive a polymer solution or polymer melt out of the spinneret.

http://dx.doi.org/10.5772/65836

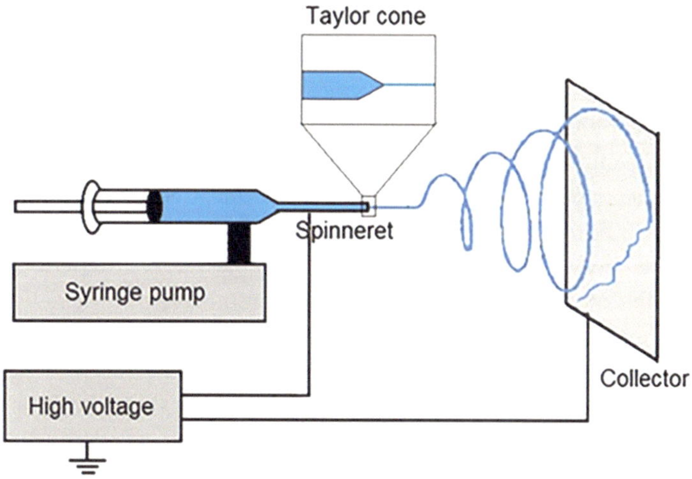

Figure 1. Schematic of a typical electrospinning system. A high voltage applied between the spinneret tip and the collector creates an electrified fluid jet, which is accelerated toward the grounded collector, forming continuous fibers.

In electrospinning, polymer fibers are formed by the generation and elongation of an electrified fluid jet. Under the application of a high voltage, when the electrostatic force from the repulsion of like charges in the fluid overcomes the surface tension, the fluid droplet coming out of the spinneret deforms to a conical shape called a Taylor cone, named after Sir Geoffrey Ingram Taylor with his pioneering work on electrified fluids [5].

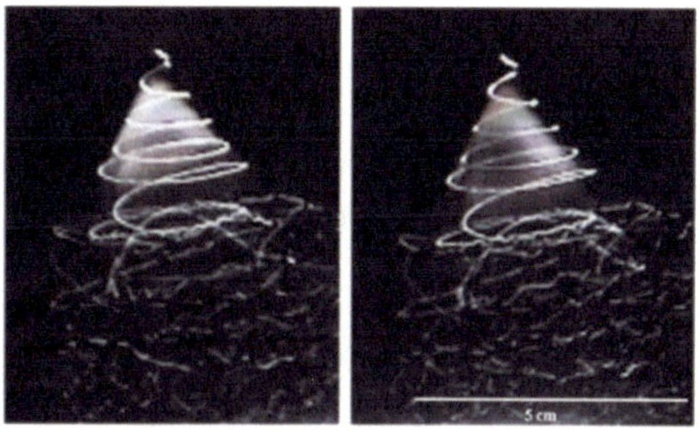

Figure 2. A pair of stereographic stroboscopic picture of a jet of polyethylene oxide, molecular weight 400,000, dissolved at a concentration of 6% in a mixture of 75% water and 25% ethanol, recorded during electrospinning. Reprinted with permission from Reneker and Yarin [13].

Taylor also proposed a "leaky dielectric" model for the moving electrified fluid that behaves like neither a perfect dielectric nor a perfect conductor [4]. The flow of charge through the fluid leads to elongation and thinning of the jet. Repulsive interactions between like charges in the fluid cause bending instability [9]. Theoretical modeling and video observation using high speed CCD cameras both showed that the fluid jet bent and stretched in a conical envelop, creating a highly spiral jet path, as shown in **Figure 2**. When the jet reaches the collector, continuous fibers are produced. The diameter of electrospun fibers usually falls in the range between tens of nm and 1 mm.

Despite the simple experimental setup, the electrospinning process is affected by many variables such as the polymer type and molecular weight, solution concentration, viscosity, surface tension and conductivity, solvent type, voltage, flow rate, distance between needle tip to collector, and ambient parameters (temperature, humidity). Several theoretical models have been developed with limited success in predicting the electrospun fiber properties [14, 15]. Obtaining desirable fiber size and distribution remains largely dependent on empirical observations. On the other hand, while a universal model is not available to predict the electrospun fiber morphology for different polymer/solvent systems under different processing conductions, some general relationships can be drawn to guide the electrospinning process, as shown in **Table 1**. The sections below discuss some of these key parameters that affect the electrospinning process.

Parameters	Effect on fiber morphology
Solution parameters	
Solvent vapor pressure	Higher porosity with higher volatility
Polymer concentration	Increasing fiber diameter with higher concentration (within optimal range)
Solution viscosity	Increasing fiber diameter with higher viscosity (within optimal range)
Solution surface tension	No conclusive link
Solution conductivity	Decreasing diameter with higher conductivity
Processing parameters	
Voltage	No conclusive link between fiber diameter and voltage; higher probability of bead formation with higher voltage
Flow rate	Increasing fiber diameter and bead formation with higher rate (above minimum rate)
Needle-collector distance	Decreasing fiber diameter with larger distance (within optimal range)
Ambient parameters	
Temperature	Decreasing fiber diameter with higher temperature
Humidity	Higher humidity induces circular pores

Table 1. General relationships between electrospinning parameters on fiber morphology, adapted with permission from Bhardwaj, Sill and Pham [16–18].

2.2. Solution parameters

A number of solution parameters (such as solvent type, polymer concentration, solution viscosity, surface tension, and conductivity) play an important role in the electrospun fiber formation and morphology. These parameters are usually correlated to each other. For example, a higher polymer concentration corresponds to a higher solution viscosity. The surface tension and conductivity are both directly related to the type of solvent used.

2.2.1. Solvent properties

The first step in electrospinning is to dissolve the polymer in a suitable solvent. As the electrified fluid jet travels toward the collector, the solvent evaporates and phase separation occurs, leading to the formation of fibers. **Table 2** lists some commonly used solvents and their property data.

	Density (g cm^{-3})	Vapor pressure (kPa at 25°C)	Boiling point (°C)	Dielectric constant	Viscosity (cP at 25°C)
Acetone	0.785	30.6	56.1	20.7	0.31
Acetic acid	1.049	2.1	118–119	6.2	1.16
Chloroform	1.49	25.9	61.2	4.8	0.53
Carbon disulfide	1.266	48.1	46.2	2.6	0.36
Cyclohexane	0.778	13.1	80.7	2.02	0.98
Dimethyl formaldehyde	0.948	0.52	152–154	36.7	0.80
Ethanol	0.789	7.87	78.4	24.5	1.07
Hexafluoro-2-propanol	1.596	16	58.2	16.7	1.02
Methanol	0.792	17.0	64.7	32.7	0.544
Tetrahydrofuran	0.889	15.2	66	7.6	0.48
Toluene	0.87	3.8	111	2.4	0.56
Water	1	3.17	100	80.1	0.89

Table 2. Properties of commonly used solvents in electrospinning.

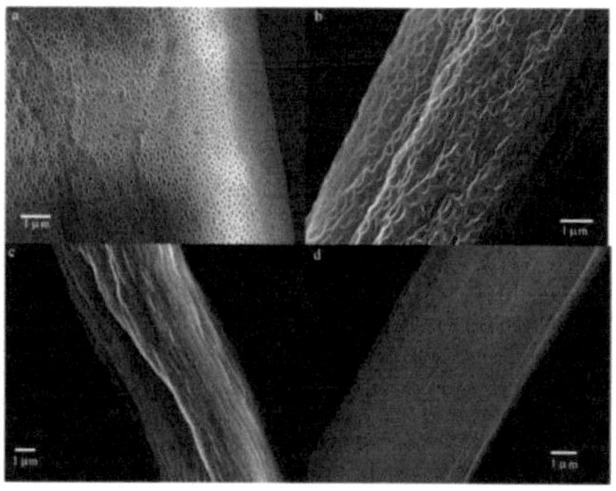

Figure 3. Field emission scanning electron microscopy images of electrospun PS fibers as a function of solution volatility: (a) 100% THF (15 kV, WD 8.9 mm); (b) 75/25% THF/DMF (15 kV, WD 9.3 mm); (c) 50/50% THF/DMF (15 kV, WD 8.8 mm); (d) 100% DMF (15 kV, WD 9.0 mm). Reprinted with permission from Megelski et al. [18].

The volatility of the solvent plays an important role in determining the fiber morphology. For example, polystyrene (PS) fibers electrospun from 100% tetrahydrofuran (THF) (more volatile) are highly porous (**Figure 3a**), whereas using 100% dimethyl formaldehyde (DMF) (less volatile) gives very smooth fibers and complete disappearance of microtexture and nanopores (**Figure 3d**). When a mixture of THF and DMF is used as the solvent, the fibers show increased pore size and decreased porosity as the solvent volatility decreases [19].

2.2.2. Solution concentration

The polymer concentration influences both the viscosity and surface tension of the solution, both of which can affect the electrospinning process. The critical entanglement concentration (**Figure 4**) is defined as the minimum concentration below which electrospray beads instead of electrospun fibers will form [20]. On the other hand, if the solution is too concentrated, the high viscosity will also inhibit the flow of fluid to the needle tip and consequently slows the electrospinning process.

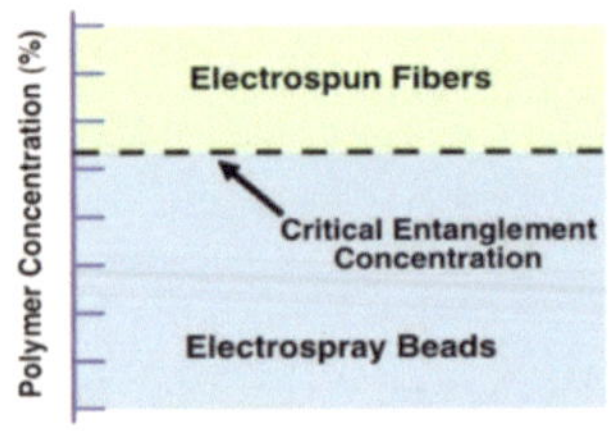

Figure 4. Critical entanglement concentration for electrospinning. Reprinted with permission from Leach et al. [20].

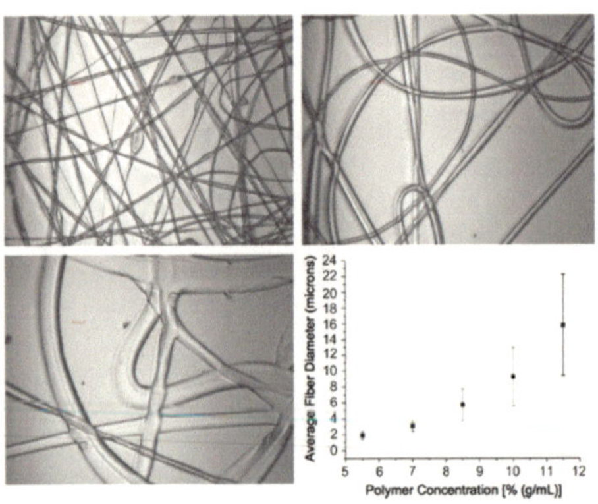

Figure 5. Effect of polymer concentration on fiber diameter. Fibers were electrospun from solutions containing varying concentrations of poly(ethylene-co-vinyl alcohol) in 70:30 (v:v) 2-propanol: DI water. Top left: fibers electrospun from a 5.5% (g/mL) solution. Top right: fibers electrospun from an 8.5% (g/mL) solution. Bottom left: fibers electrospun from an 11.5% (g/mL) solution. The following processing parameters were used for all experiments: applied voltage: 20 kV, flow rate: 3 mL/h, capillary-collector distance: approximately 25 cm. In the bottom right panel the relationship between the average fiber diameter and the polymer concentration is given. Note that the mean fiber diameter increases monotonically with increasing polymer concentration. Additionally, it is evident that ribbon-like fibers are formed at higher concentrations (11.5%), which indicates incomplete polymer drying. (Error bars represent the standard deviation.). Reprinted with permission from Sill and von Recum [18].

The optimal range of solution concentration varies for different polymer/solvent systems. For example, for an aqueous solution of polyethylene oxide (MW 400,000), fibers can be produced when the solution concentration is in the range of 4–10 wt% [21]. Within the optimal range, the fiber diameter usually increases with the increasing concentration, as shown in **Figure 5**.

2.3. Processing parameters

For a specific polymer/solvent system, there is usually an optimal range of processing parameters to obtain electrospun fibers with desirable size and distribution, as summarized in **Table 1**.

2.3.1. Applied voltage

The applied voltage is an important parameter in the electrospinning process. A threshold voltage needs to be reached to initiate the fiber formation process. The relationship between the applied voltage and fiber diameter is not clear. Some studies reported a reduction in the fiber diameter with increasing voltage, presumably due to greater stretching of the fluid jet by the stronger electric field [22]. Other studies showed that higher voltage led to increase in fiber diameter [23]. In addition, higher voltage has also been found to facilitate the bead formation [21].

2.3.2. Collector configuration

Earlier electrospinning experiments generally used a stationary collector to obtain randomly oriented fibers. As aligned electrospun fibers were demonstrated to be more important for many applications ranging from tissue engineered scaffold construction to photonics and fuel cells, new collectors have been developed to better control the fiber orientation, as illustrated in **Figure 6**. **Figure 7** shows randomly oriented fiber mat and aligned fibers by changing the collector configuration.

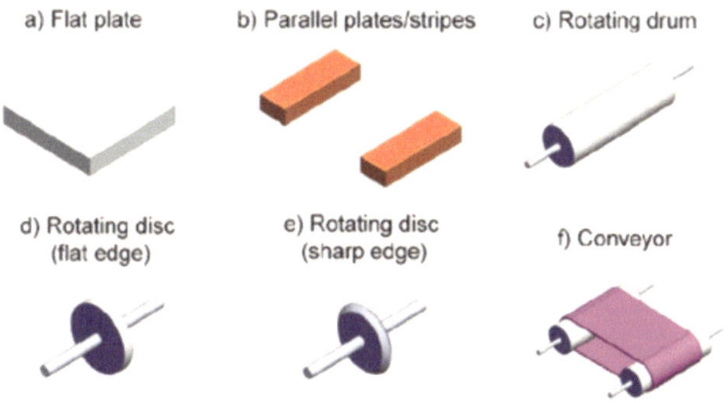

Figure 6. Sketches of electrospinning collectors (a) is used to collect randomly oriented fibers; (b)–(f) are used to collect aligned fibers. Reprinted with permission from Persano et al. [25].

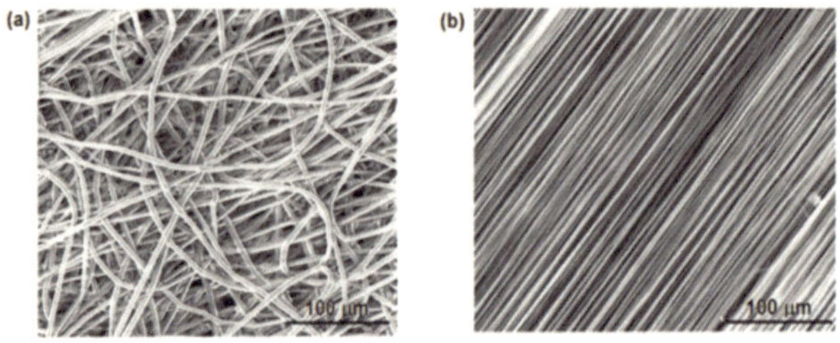

Figure 7. Electrospun PCL fibers (20% PCL, 25 kV) collected as (a) a randomly oriented fiber mat using a stationary collector; (b) aligned fibers using a rotating disc with sharp edge.

More recent collector development has made it possible to create complex bi- and tri-dimensional architectures in a single run. Careful design of the 3D collectors with static collection allows one-step production of micro- and macro patterned tubes (diameter 500–5 mm) with different shapes and architectures, and T-shaped interconnections [24]. Such structures are attractive for the regeneration of many tubular fibrous tissues (**Figure 8**).

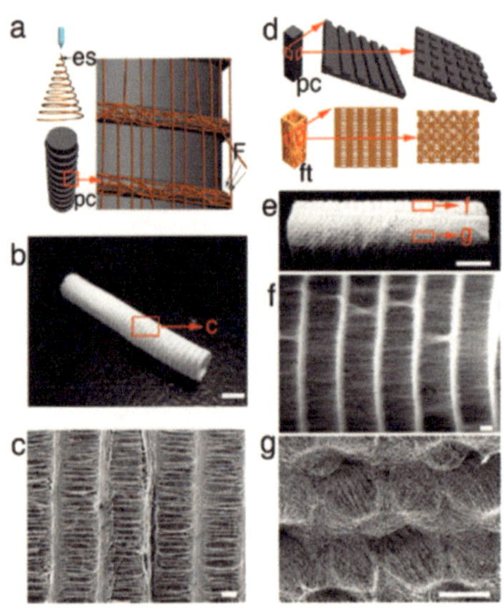

Figure 8. (a) Schematic illustration of collecting process using a cylindrical collector with equally spaced circular protrusions (es, electrospinning process; pc, patterned collector). (b) A fibrous tube with patterned architectures (scale bar, 5 mm). (c) Magnified image of panel b (scale bar, 200 μm). (d) Schematic illustration of collectors with two different patterns and relevant fibrous tube (pc, patterned collector; ft, fibrous tube). (e) A fibrous tube with two different patterns (scale bar, 5 mm). (f) and (g) Magnified images of two different patterns of panel e (scale bar 200 μm). Reprinted with permission from Zhang and Chang [23].

2.3.3. *Spinneret design*

A single needle attached to a syringe pump is the most commonly used spinneret design in electrospinning. Other designs have been developed to either create a core/shell fiber structure (**Figure 9d** and **e**) or to increase the throughput and thickness of the fiber production (**Figure 9b** and **c**). The core/shell structure (**Figure 10**) can be utilized to synthesize a composite electrospun fibrous scaffold [27, 28]. Therapeutic agents and signaling molecules can also be incorporated into the core or shell of electrospun fibers with tailored release profile to modulate the cell behavior.

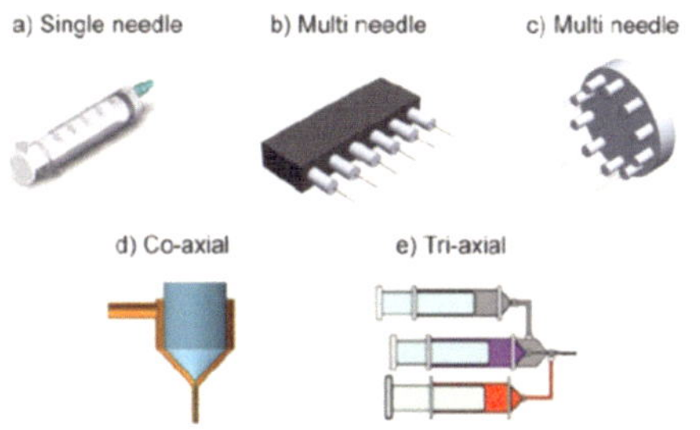

Figure 9. Sketches of different spinneret designs used in electrospinning. (a) single needle; (b) Multi needle with linear configuration; (c) Multi needle with circular configuration; (d) Co-axial needle; and (e) tri-axial needle. Reprinted with permission from Persano [3].

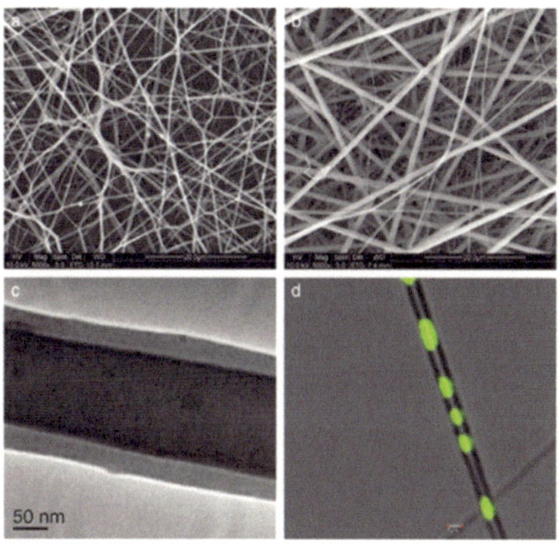

Figure 10. FESEM images showing the fiber morphology of (a) fibrinogen nanofibers, (b) PGS/fibrinogen core/shell fibers, (c) PGS/fibrinogen core/shell morphology (higher magnification), and (d) PGS/fibrinogen core/shell (FITC dye in core). Reprinted with permission from Ravichandran et al. [26].

2.4. Ambient parameters

Ambient parameters such as temperature and humidity have also been found to affect the fiber morphology. As most solution viscosity changes inversely with temperature, higher temperature generally leads to thinner fibers, as has been supported by Mit-upathum's study [29].

Higher humidity has been found to induce the appearance of circular pores on electrospun fibers. When the humidity is too low, the fast drying of solvent may cause clogging of the needle tip.

3. Surface modification of electrospun fibers

3.1. Electrospun fibers as biomimetic tissue engineering scaffolds

Tissue engineering represents a revolutionary approach to the repair and regeneration of diseased tissues and organs through a combination of biomaterial scaffolds, cells, and regulators. A biomimetic approach is often used in tissue engineering scaffold design that aims to mimic certain advantageous features of the extracellular matrix (ECM), such as the materials' composition, surface chemistry, mechanical properties, structural features, and growth factor delivery strategies [30]. Electrospinning is an attractive method to produce biomimetic tissue engineering scaffolds due to the simple and low-cost experimental setup, the wide variety of materials that can be electrospun, and inherent nanostructure feature with large surface area and high porosity.

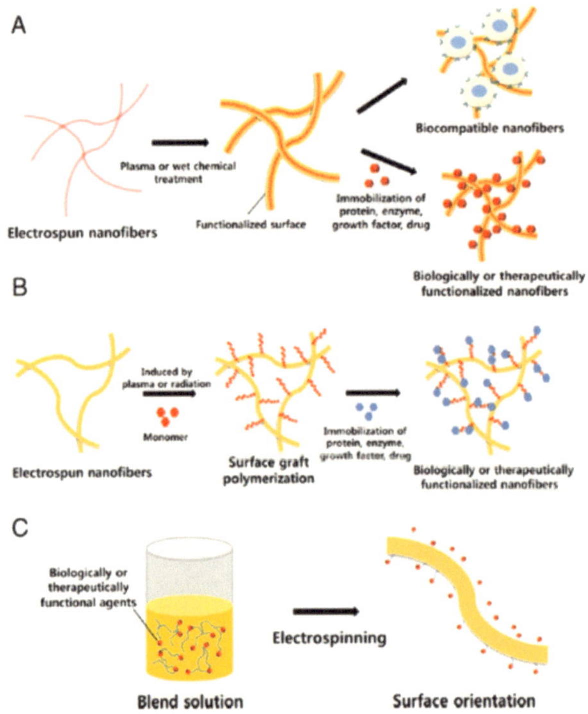

Figure 11. Surface modification methods for electrospun polymer fibers. (A) Plasma or wet chemical treatment; (B) surface graft polymerization; and (C) co-electrospinning. Reprinted with permission from [5].

Electrospun fibers from natural polymers (such as collagen, silk, chitin, elastin, and fibrinogen) usually exhibit similar structural features as the ECM, making it easy to guide or direct cellular response during the tissue regeneration process. Major limitations of these natural polymers are the difficulty to create fibers reproducibly due to large variations in the structure and properties of the source polymers, and their generally insufficient mechanical properties. Electrospun synthetic polymers usually have higher mechanical strength than natural ones. Their structure and properties are also more reproducible and tailorable. On the other hand, many electrospun synthetic polymer fibers are hydrophobic and lack bioactive components on the surface to modulate cell response.

Surface modification is a cost effective approach to change the surface properties of the material without significantly altering its bulk properties. A common strategy in tissue engineering scaffold design is to incorporate bioactive molecules onto the fiber surface to improve biocompatibility or induce specific cellular response such as cell adhesion, proliferation, antigen presentation, and activation of explicit pathways to promote specific cell functions.

In general, any method used for surface modification of bulk polymers can be applied for electrospun fiber functionalization as long as it does not alter the fibrous structure. As illustrated in **Figure 11** and **Table 3**, the electrospun polymer fiber surface can be modified through physical entrapment, plasma treatment, chemical immobilization, or coelectrospinning of bioactive molecules.

Method	Electrospun fibers	Immobilized agents	TE applications
Physical entrapment	PCL	Gelatin, calcium phosphate, soluble eggshell membrane protein	Bone, blood vessel, general
	P(LLA-CL)	Collagen	Blood vessel
	PLLA	Laminin	Neural
	Silk	Fibrin	Cartilage
Wet chemical (NaOH hydrolysis)	PLLA	HAp	Bone
	PCL	–	Bone, ligament
Chemical immobilization	PET/PMAA	Gelatin	Blood vessel
	PLLA (PGA, PLGA) /PAA		General
	PCL	RGD	Ligament
Co-electrospinning	PLLA	HAp	Bone
	PLGA-PEG-NH_2	RGD	General

Abbreviations: PCL, poly(ε-caprolactone); P(LLA-CL), poly(L-lactic acid)-co-poly(ε-caprolactone); PLLA, poly(L-lactic acid); PLLC, poly(L-lactide-co-caprolactone) copolymer; PET, polyethylene terephthalate; PGA, poly(glycolic acid); PLGA, poly(lactic-co-glycolic acid); PMAA, poly(methacrylic acid); PAA, poly(acrylic acid); PVP, poly(4-vinylpyridine); HAp, hydroxyapatite; PEG, poly(ethylene glycol); RGD, Arg-Gly-Asp.

Table 3. Surface modification techniques for electrospun polymer fibers in tissue engineering applications, adapted with permission from Yoo et al. [31].

3.2. Physical entrapment

Soaking the electrospun fibers in a solution that contains a high concentration of the bioactive components is the simplest surface modification approach. The high porosity of the nano-structure facilitates the adsorption of bioactive molecules through hydrogen bonding, van der Waals force, or electrostatic forces. For example, electrospun chitosan nanofibers were surface modified with the fibronectin, an adhesion molecule, to enhance the attachment of rat cardiomyocytes onto the 3-D cardiac tissue scaffold [32].

During the last decade our knowledge about amphiphilic proteins like hydrophobias increased tremendously and these small proteins were used to modify the surface of biosensors to render them more hydrophilic proteins [33]. The mechanism of their action as described by several investigators is that they self-assemble into amphiphilic membranes leading to increase wettability of hydrophobic surfaces of materials [19, 34]. This specific nature attracted more investigators to use them to incorporate growth factors and other proteins to improve cellular attachment to materials with good mechanical properties but with poor biological activity. Zhao et al. demonstrated that the hydrophilicity of electrospun PCL fibers can be improved by using a fusion protein consisting of HGFI and vascular endothelial growth factor (VEGF). In this study, the self-assembled layer of VEGF-HGFI effectively enhanced the adhesion, migration, and proliferation of human umbilical vein endothelial cells [35]. Another polymer that is particularly useful for coating and modification of various surfaces is Polydopamine (PDA) formed by the oxidation of dopamine. This polymer was used to modify the surface of PLA nanofibers in order to promote the adhesion and proliferation of human adipose-derived stem cells (hADSCs), cell attachment was significantly enhanced on PDA/PLA modified surfaces relative to control [28].

3.3. Plasma treatment

Synthetic polymers are generally hydrophobic. Plasma treatment using oxygen or air can introduce hydroxyl groups onto the fiber surface and effectively decrease its hydrophobicity. Several studies have reported increased hydrophilicity and enhanced attachment of fibroblasts on electrospun PCL or poly(butylene carbonate) fibers [36]. Plasma treatment using oxygen, ammonia, or air generates carboxyl groups, or amine groups on the surface, thus serving as a precursor to other surface modification methods. For example, oxygen plasma treatment of PLLA nanofibers followed by cross-linking with cathionized gelatin allowed better chondrocyte attachment to the functionalized PLLA nanofibers [11].

3.4. Wet chemical treatment

Acidic or alkaline solutions can be used to induce partial surface hydrolysis of electrospun polyester fibers to modify the surface wettability or to create nanotopography.

The random chemical scission of ester links leads to generation of carboxylic and hydroxyl groups. For example, NaOH-treated PLLA nanofibrous mesh showed greatly enhanced nucleation and growth of hydroxyapatite minerals on the fiber surface, likely due to the chelation of calcium ions by carboxylic acids after the NaOH treatment. When electrospun

PCL nanofibers were treated with 5 M NaOH, the fiber wettability was dramatically enhanced with almost zero water contact angle due to the capillary action on the highly rough surface. The nanotopography also led to more favorable cell adhesion.

3.5. Chemical immobilization

Compared to physical entrapment, chemical immobilization provides a stronger covalent attachment of bioactive molecules on the electrospun fiber surface. The fibers need to be pretreated to generate reactive functional groups before bioactive molecules can be immobilized, as illustrated in **Figure 12**.

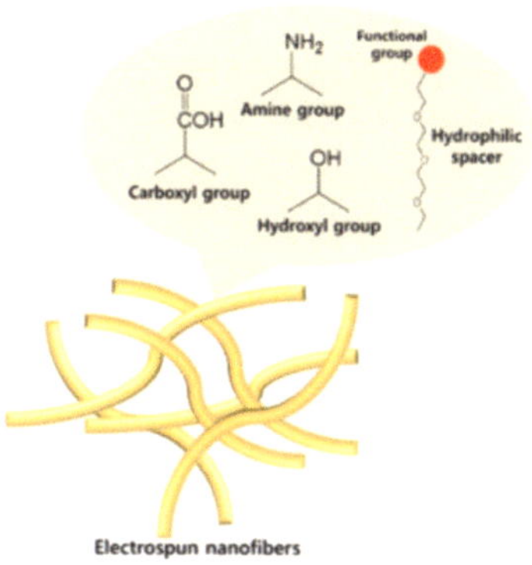

Figure 12. Common surface functional groups for immobilization of bioactive molecules on electrospun fibers. Reprinted with permission from Yoo et al. [31].

A most commonly used chemical immobilization approach involves the use of 1-ethyl-3-(3-dimethylaminopropyl) carbodiimide (EDC) and *N*-hydroxysuccinimide (NHS) to activate carboxylic acid groups on the fiber surface, which are then conjugated to primary amine groups of bioactive molecules. In the case of the PCL, this step can be accomplished by alkaline hydrolysis using sodium hydroxide surface treatment [37]. A study by Cheng showed that covalent immobilization via EDC coupling of collagen onto electrospun nanofiber matrices of PCL-chitosan (CS) blends significantly improved rat bone marrow derived stromal cells (rBMSCs) adhesion, spreading, proliferation, and osteogenic differentiation as compared to control groups [38]. Other cross-linking agents are aldehydes like gluteraldehyde are used where maintenance of structural rigidity of protein is required. This method was used to cross-link hydroxyethyl cellulose nanofibrous mats to improve their cellular adhesion characteristics and stability for potential scaffold for skin tissue engineering [39]. Gluteraldehyde was also used by Krishnan and colleagues for cross-linking Xylan, a natural polysaccharide and polyvinyl alcohol (PVA) to produce Xylan/PVA nanofibers for skin tissue engineering [40].

3.6. Co-electrospinning

The surface modification methods discussed above all use posttreatment on electrospun fibers. Co-electrospinning the bioactive agents with polymers provides an alternative approach to biofunctionalize the fiber *in situ*. The bioactive agents can be directly mixed with the polymer solution before electrospinning. The core/shell structure discussed in Section 2.3.3 can also be utilized to directly incorporate bioactive molecules on the fiber surface.

4. Application of electrospun fibers in tissue engineering

The ease of producing nanofibers and the variety of biocompatible polymers that can be formed by electrospinning have uncovered many of their potential applications in emerging fields such as tissue engineering. One of the many advantages of electrospun nanofiber scaffolds is that their surface can be modified by controlling the electrospinning parameters to obtain the topography that best fits the application. Another advantage is that nanofiber sheets or matrices can be formed into almost any shape (patches, mats, tubes, fibers, multilayered matrices) based on the site of desired implantation. As shown in **Table 4**, electrospun fibrous scaffolds have been used to regenerate a variety of tissues such as the skin, vasculature, neural, bone, ligament, and tendon.

No.	Polymer	Solvent	Fiber diameter	Nozzle configuration	Application (cell type/drug)
General tissue engineering (T.E.)					
1	Poly((-caprolactone)	(a) Chloroform and methanol	2–10 μm	Single nozzle	General T.E. (rat marrow stromal cells)
		(b) Chloroform and DMF	~600 nm		
2	(a) Poly(∈-caprolactone) (core)	(a) Chloroform and DMF	500–900 nm	Coaxial	General T.E. (none used)
	(b) Zein (shell)	(b) DMF			
3	(a) Poly(∈-caprolactone) (cone)	(a) 2,2,2-Trifluoroethanol	~513 nm	Coaxial	General T.E. (human dermal fibroblasts)
	(b) Collagen (shell)	(b) 2,2,2-Trifluoroethanol			
4	Poly(D,L-lactic-*co*-glycolic acid) and PLGA-b-PEG-NH_2	DMF and THF	449–1312 nm	Single nozzle	General T.E. (NIH3T3 fibroblasts)
5	Poly(D,L-lactide-*co*-glycolide)	DMF AND THF	500–800 nm	Single nozzle	General T.E. (human mesenchymal stem cells and BALB/c C7 mouse fibroblasts)
6	Poly(ethylene glycol-*co*-lactide)	DMF and acetone	1.25–4.25 μm	Single nozzle	General T.E. (none used)

No.	Polymer	Solvent	Fiber diameter	Nozzle configuration	Application (cell type/drug)
7	Poly(ethylene-*co* -vinyl alcohol)	2-Propanol and water	0.2–8.0 μm	Single nozzle	General T.E. (human aortic smooth muscle cells and human dermal fibroblasts)
8	Collagen	HFP	180–250 nm	Single nozzle	General T.E. (rabbit conjunctiva fibroblasts)
9	Collagen	HFP	100–730 nm	Single nozzle	General T.E. (aortic smooth muscle cells)
10	(a) Collagen	(a) HFP and acetic acid	3–6 μm	Single nozzle	General T.E. (human osteosarcoma cells)
	(b) Gelatin	(b) HFP	2–6 μm		
11	Gelatin	2,2,2-Trifluoroethanol	0.29–9.10 μm	Single nozzle	General T.E. (none used)
12	Fibrinogen	HFP and 10× minimal essential medium	0.12–0.61 μm	Single nozzle	General T.E. (neonatal rat cardiac fibroblasts)
13	Poly(glycolic acid) and chitin	HFP	130–380 nm	Single nozzle	General T.E. (normal human epidermal fibroblasts)
14	Collagen and Poly(ethylene oxide)	10 mM HCl (pH 2.0)	100–150 nm	Single nozzle	General T.E. (none used)
15	Poly(DTE carbonate)	DCM	1.9–5.8 μm	Single nozzle	General T.E. (NIH3T3) (mouse embryo fibroblasts), MCF-7 (human mammary carcinoma), PC-12 (rat adrenal pheochromocytoma) and KB (KB/HeLa; human cervical carcinoma))
Vascular T.E.					
1	Poly(∈-caprolactone)	Chloroform and DMF	0.2–1 μm	Single nozzle	Vascular T.E. (human coronary artery endothelial cells)
2	Poly(D,L-lactide-*co*-glycolide), collagen, and elastin	HFP	720 ± 350 nm	Single nozzle	Vascular T.E. (bovine endothelial and smooth muscle cells)
3	Poly(L-lactide-*co*-∈-caprolactone)	Acetone	200–800 nm	Single nozzle	Vascular T.E. (human coronary artery smooth muscle cells)
4	Poly(L-lactide-*co*-∈-caprolactone)	HFP	799–820 nm	Single nozzle	Vascular T.E. (human umbilical vein endothelial cells (HUVECs))

No.	Polymer	Solvent	Fiber diameter	Nozzle configuration	Application (cell type/drug)
5	Poly(L-lactide-*co*-∈caprolactone)	HFP	700–800 nm	Single nozzle	Vascular T.E. (none used)
6	Poly(L-lactide-*co*-∈-caprolactone)	HFP	100–300 nm	Single nozzle	Vascular T.E. (human coronary artery endothelial cells (HCAECs))
7	Poly(propylene carbonate)	Chloroform	~5 μm	Single nozzle	Vascular T.E. (rat bone marrow mesenchymal stem cells)
Neural T.E.					
1	(a) Poly(∈-caprolactone)	(a) Chloroform and methanol	559 ± 300 nm	Single nozzle	Neural T.E. (DRG explants, dissociated DRG, Schwann cells, olfactory ensheathing cells, and fibroblasts)
2	Poly(L-lactic acid)	DMF and DCM	300–3500 nm	Single nozzle	Neural T.E. (mouse neural stem cells)
Bone T.E.					
1	Poly(L-lactic acid) and hydroxylapatite	DCM and 1,4-dioxane	~313 nm	Single nozzle	Bone T.E. (MG-63 osteoblasts)
Skin T.E.					
1	Chitin	HFP	0.163–8.77 μm	Single nozzle	Skin T.E. (normal human oral keratinocytes, normal human epidermal keratinocytes, and normal human gingival fibroblasts)

Table 4. Example electropsun polymer fiber scaffolds used in tissue engineering, adapted with permission from Sill and von Recum [18].

4.1. Skin tissue regeneration

Flat and flexible nanofiber sheets are preferred to promote skin healing, regeneration, and substitution with the advantage of providing coverage of the exposed dermis in most situations. One study describes the potential use of polycaprolactone (PCL) and collagen nanofiber matrices as a dermal substitute. They performed *in vitro* testing that showed matrices supported the attachment and proliferation of human dermal fibroblasts as compared to the control [41]. Another study by Veleirinho and colleagues evaluated hybrid nanofibrous mats prepared by electrospinning consisting of poly(3-hydroxybutyrate-co-3-hydroxyvalerate) (PHBV) and chitosan for wound healing using a full thickness wound healing model. They showed that these scaffolds promoted wound healing in the rat [42]. The positive feedback from the *in vitro* and the animal studies stimulated the movement toward human clinical trials including the use of nanofibers for drug release to promote healing. A double blind, randomized, placebo controlled clinical trial was performed in 2007and completed in 2011, to evaluate the effectiveness and safety of a novel nitric oxide (NO) releasing wound dressing patch (PATHON) for the treatment of diabetic foot ulcers. The patch consists of electrospun nanofiber mesh that encapsulates a NO donor, permitting a constant release of NO for a 12 h period. Up to now, there are no available results [43]. Some preliminary results are published by Demircan and colleagues on the clinical usage of the collagen-elastin matrix in 15 children with facial burns. They show encouraging early results: the graft quality was close to normal skin in terms of vascularity, elasticity, pliability, texture, and color [44].

4.2. Cardiovascular tissue engineering

The feasibility of forming various construct from electrospun nanofiber allowed the use of sheets or tubes for cardiovascular tissue engineering applications. In an experiment to study the effect of nanofibrous collagen/elastin/polycaprolactone patch loaded with cardiac nature protein (NP) on cardiac repair after myocardial infarction, it was found that bone-marrow cells seeded sheets improved the cardiac function in MI mice after 4 weeks of transplantation [45]. Due to the ability to modify the mechanical and chemical properties of nanofiber material, and in order to improve the patency rate following small-diameter vascular grafting, several groups are investigating the potential use of tubular nanofibers for small blood vessel substitution. One group was testing heparin-bonded P(LLA-CL) vascular scaffolds seeded with autologous endothelial cells EC and implanted into canine arterial model to promote graft patency rate. The scaffold inner layer was fabricated by heparin-bonded P(LLA-CL) nanofibers through coaxial electrospinning, while the outer layer was woven by pure P(LLA-CL) nanofibers. In this *in vivo* canine femoral artery replacement study the authors were able to demonstrate that the proposed biomaterial significantly promoted the 24 weeks patency rate (88.9%) in the experiment group as compared to control (12.5%) [46].

4.3. Bone tissue engineering

Tissue engineered bone materials are attractive alternative to synthetic grafts since they are biocompatible, bioactive, and designed to degrade after the appropriate cells start making their own matrix. Collagen is extensively involved in the process of adhesion and proliferation of many cell types and its physicochemical properties can be readily modified by cross-linking with many reagents since it has amino, carboxyl and hydroxyl groups that can serve as cross-linking sites. On the other hand stem cells are becoming more involved in tissue regeneration. Thus, the combination of collagen nanofibers and stem cells make them suitable for the engineering of various tissues including bone. Cheng et al. reported that PCL electrospun nanofiber matrices blended with chitosan and functionalize with type I collagen regulated the differentiation of rat bone marrow derived stromal cells (rBMSCs) into osteogenic lineage [47].

In our laboratories we fabricated aligned electrospun collagen nanofibers, we cross-linked them with EDC and we seeded them with rBMSCs (**Figure 13**). In this work, we tested the differentiation capability of these cells on this matrix without the use of any differentiation factors. After 5 weeks of culture, the BMSC differentiated into osteoblasts and were able to deposit calcium minerals on the material surface (**Figure 14**) [48].

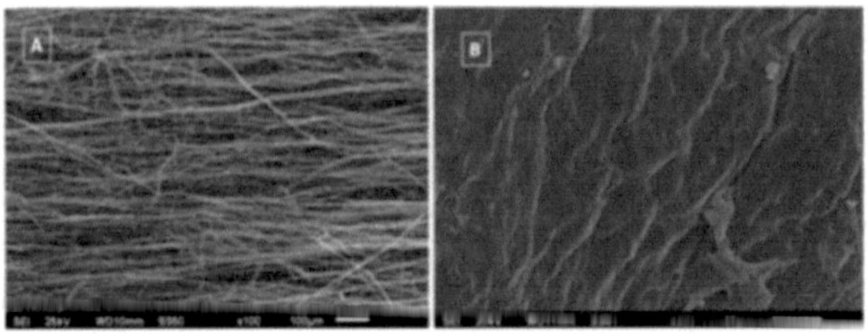

Figure 13. (A) SEM image showing the morphology of collagen nanofibers before seeding. (B) SEM image of seeded nanofibers after 5 weeks of culture showing complete covering of the material by differentiating cells.

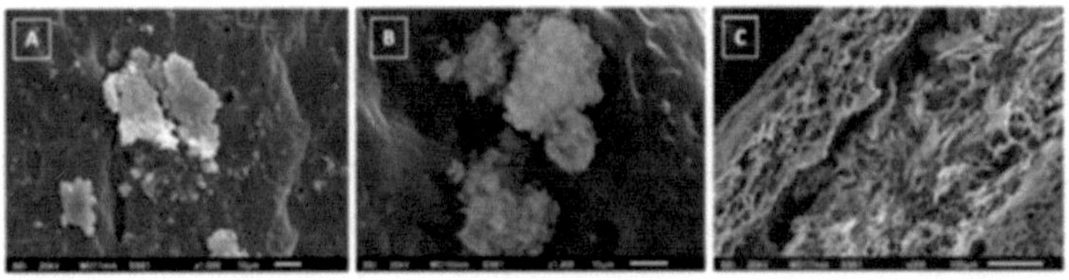

Figure 14. SEM images of nanoscaffolds after 5 weeks of culture with BMSCs showing different morphology of the mineral deposition on the surface of the material, (A and B) and a cross-section view showing the porous structure of the material (C).

4.4. Ligament and tendon tissue engineering

The ability to regenerate tendon or ligament tissues is a major goal of tissue engineering. The specific structure and alignment of collagen fibers within a tendon plays a significant role in their tensile behavior. They preferentially align to the applied stresses. Many of the used biomaterials so far support the cellular growth and function but lack the initial mechanical strengths needed before the new tissue takes place. A recent work by Wang et al. explored the effect of aligned nanofibers on inducing tenogenic phenotype of human dermal fibroblasts (HDFs) *in vitro* and on inducing tendon regeneration *in vivo*. They showed that the aligned nanofibers induced tenogenic phenotype and Achilles tendon regeneration in a rat model [49]. An interesting application of PCL nanofibers was introduced by Martin and colleagues that involves the modification of PCL with radio opaque particles of zirconia to enhance the visualization of the implant by X-ray. The modified material was seeded with bovine MSCs and implanted *in vivo* in a model of total disc replacement in the rat coccygeal spine for 4 weeks. In this study the authors showed that radio opaque nanoparticles can be included into nanofibrous scaffolds and thus providing a biocompatible template for *in vivo* visualization, future image-guided implantation and long term evaluation of scaffold location and performance [50].

Our lab has developed a braiding technique for electrospun PCL fibers to mimic the hierarchical architecture of collagen seen in native anterior cruciate ligament (ACL). The braided scaffolds also showed comparable mechanical properties as the native ACL (**Figure 15**, unpublished). Ongoing work is focused on evaluating the biocompatibility of the braided scaffolds.

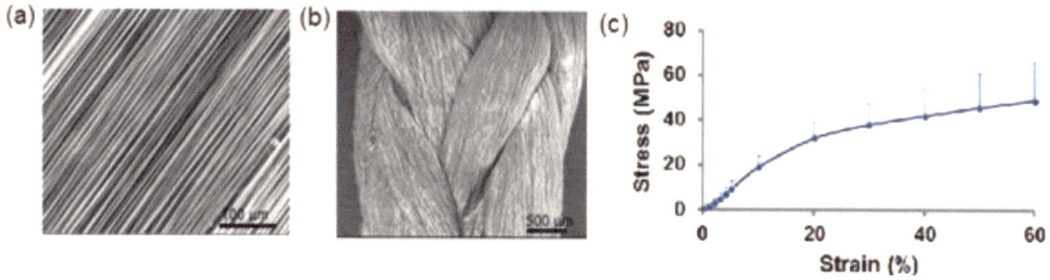

Figure 15. (a) ESEM image of electrospun PCL nanofibers; (b) a braided structure using nine fiber strips by first braiding three fiber strips into a triple helix, then braiding three such subunits into a second-order triple helix structure; and (c) stress-strain plot of 11 samples from 6 rounds of braids giving an elastic modulus (228 ± 50 MPa), ultimate tensile strength (52 ± 17 MPa) and toe region (3.3 ± 1.6%).

5. Summary

The recent interest in electrospinning as a tissue engineering scaffold fabrication method is attributed to its relatively simple and inexpensive experimental setup, the wide variety of applicable materials, and the inherent nanoscale nature of the fibers closely mimicking that of the ECM. Further study of the electrospinning process and new development in the spinneret and needle configuration, along with effective utilization of surface modification methods, will help to create more biomimetic scaffolds tailored to the specific tissue regeneration applications.

Author details

Yawen Li[1*] and Therese Bou-Akl[2]

*Address all correspondence to: yli@ltu.edu

1 Lawrence Technological University, Southfield, MI, USA

2 St John Providence Hospital Orthopedic Research Laboratory, Southfield, MI, USA

References

[1] Gilbert, W., *De Magnete, Magneticisque Corporibus, et de Magno Magnete Tellure (On the Magnet and Magnetic Bodies, and on That Great Magnet the Earth)*, in *Wikipedia: The free encyclopedia. (2004, July 22)*. FL: Wikimedia Foundation, Inc. Retrieved August 10, 2004, from http://www.wikipedia.org. 1628, Peter Short: London.

[2] Zeleny, J., *The electrical discharge from liquid points, and a hydrostatic method of measuring the electric intensity at their surfaces.* Phys Rev, 1914. 3(2): pp. 69–91.

[3] Zeleny, J., *Instability of electrified liquid surfaces.* Phys Rev, 1917. 10(1): pp. 1–6.

[4] Taylor, G., *Electrically driven jets.* Proceedings of the Royal Society of London. A: Mathematical, Physical & Engineering Sciences, 1969. 313: pp. 453–475.

[5] Taylor, G., *Disintegration of water droplets in an electric field.* Proceedings of the Royal Society. 1964. A 280 (1382): pp. 383–397.

[6] Darrell H.R., et al., *Bending instability of electrically charged liquid jets of polymer solutions in electrospinning.* J Appl Phys, 2000. 87(4531): pp. 2494.

[7] Hohman, M.M., et al., *Electrospinning and electrically forced jets. I. Stability theory.* Phys Fluids, 2001. 13(8): p. 2201.

[8] Rutledge, G.C. and S.V. Fridrikh, *Formation of fibers by electrospinning.* Adv Drug Deliv Rev, 2007. 59(14): pp. 1384–1391.

[9] Yarin A.L., Koombhongse S., and R. DH., *Bending instability in electrospinning of nanofibers.* J Appl Phys, 2001. 89(5): pp. 3018–3026.

[10] Jayesh D. and H.R. Darrell, *Electrospinning process and applications of electrospun fibers.* J Electrost, 1995. 35(2–3): pp. 151–160.

[11] Chen, J.P. and C.H. Su, *Surface modification of electrospun PLLA nanofibers by plasma treatment and cationized gelatin immobilization for cartilage tissue engineering.* Acta Biomater, 2011. 7(1): pp. 234–43.

[12] Langer, R. and J.P. Vacanti, *Tissue engineering.* Science, 1993. 260(5110): pp. 920–926.

[13] Reneker, D.H. and A.L. Yarin, *Electrospinning jets and polymer nanofibers.* Polymer, 2008. 49(10): pp. 2387–2425.

[14] Fridrikh, S.V., et al., *Controlling the fiber diameter during electrospinning.* Phys Rev Lett, 2003. 90(14): p. 144502.

[15] Thompson, C., *An analysis of variable effects on a theoretical model of the electrospin process for making nanofibers.* Dissertation, the University of Akron, 2007.

[16] Bhardwaj, N. and S.C. Kundu, *Electrospinning: a fascinating fiber fabrication technique.* Biotechnol Adv, 2010. 28(3): pp. 325–347.

[17] Pham, Q.P., U. Sharma, and A.G. Mikos, *Electrospinning of polymeric nanofibers for tissue engineering applications: a review.* Tissue Eng, 2006. 12(5): pp. 1197–1211.

[18] Sill, T.J. and H.A. von Recum, *Electrospinning: applications in drug delivery and tissue engineering.* Biomaterials, 2008. 29(13): pp. 1989–2006.

[19] Megelski, S., et al., *Micro and nanostructured surface morphology on electrospun polymer fibers.* Macromolecules, 2002. 35(22): pp. 8456–8466.

[20] Leach, M.K., et al., *Electrospinning fundamentals: optimizing solution and apparatus parameters.* J Vis Exp, 2011.

[21] Deitzel, J.M., et al., *The effect of processing variables on the morphology of electrospun nanofibers and textiles.* Polymer, 2001. 42(1): pp. 2611272.

[22] Yuan, X., et al., *Morphology of ultrafine polysulfone fibers prepared by electrospinning.* Polym Int, 2004. 53(11): pp. 1704–1710.

[23] Zhang, C., et al., *Study on morphology of electrospun poly (vinyl alcohol) mats.* Eur Polym J, 2005. 41(3): pp. 423–432.

[24] Zhang, D. and J. Chang, *Electrospinning of three-dimensional nanofibrous tubes with controllable architectures.* Nano Lett, 2008. 8(10): pp. 3283–3287.

[25] Persano, L., et al., *Industrial upscaling of electrospinning and applications of polymer nanofibers: a review.* Macromol Mater Eng, 2013. 298(5): pp. 504–520.

[26] Ravichandran, R., et al., *Expression of cardiac proteins in neonatal cardiomyocytes on PGS/fibrinogen core/shell substrate for cardiac tissue engineering.* Int J Cardiol, 2013. 167(4): pp. 1461-1468.

[27] Song, W., et al., *Coaxial PCL/PVA electrospun nanofibers: osseointegration enhancer and controlled drug release device.* Biofabrication, 2013. 5(3): pp. 035006.

[28] Sperling, L.E., et al., *Advantages and challenges offered by biofunctional core-shell fiber systems for tissue engineering and drug delivery.* Drug Discov Today, 2016.

[29] Mit-uppatham, C., Nithitanakul, M. and P. Supaphol, *Ultrafine electrospun polyamide-6 fibers: effect of solution conditions on morphology and average fiber diameter.* Macromol Chem Phys, 2004. 205(17): pp. 2327–2338.

[30] Landis, W.J., Silver, F.H. and J.W. Freeman, *Collagen as a scaffold for biomimetic mineralization of vertebrate tissues.* J Mater Chem, 2006. 16(16): pp. 1495-1503.

[31] Yoo, H.S., Kim, T.G. and T.G. Park, *Surface-functionalized electrospun nanofibers for tissue engineering and drug delivery.* Adv Drug Deliv Rev, 2009. 61(12): pp. 1033–1042.

[32] Hussain, A., et al., *Functional 3-D cardiac co-culture model using bioactive chitosan nanofiber scaffolds.* Biotechnol Bioeng, 2013. 110(2): pp. 637–647.

[33] Hou, S., et al., *Surface modification using a novel type I hydrophobin HGFI.* Anal Bioanal Chem, 2009. 394(3): pp. 783–789.

[34] Wosten, H.A., *Hydrophobins: multipurpose proteins.* Annu Rev Microbiol, 2001. 55: pp. 625-646.

[35] Zhao, L., et al., *Functional modification of fibrous PCL scaffolds with fusion protein VEGF-HGFI enhanced cellularization and vascularization.* Adv Healthc Mater, 2016.

[36] Surucu, S. and H. Turkoglu Sasmazel, *DBD atmospheric plasma-modified, electrospun, layer-by-layer polymeric scaffolds for L929 fibroblast cell cultivation.* J Biomater Sci Polym Ed, 2016. 27(2): pp. 111–32.

[37] Rosellini, E., et al., *Surface chemical immobilization of bioactive peptides on synthetic polymers for cardiac tissue engineering.* J Biomater Sci Polym Ed, 2015. 26(9): pp. 515–533.

[38] Cheng, L., et al., *Surface biofunctional drug-loaded electrospun fibrous scaffolds for comprehensive repairing hypertrophic scars.* Biomaterials, 2016. 83: pp. 169–181.

[39] Zulkifli, F.H., et al., *Improved cellular response of chemically crosslinked collagen incorporated hydroxyethyl cellulose/poly(vinyl) alcohol nanofibers scaffold.* J Biomater Appl, 2015. 29(7): pp. 1014–1027.

[40] Krishnan, R., et al., *Polysaccharide nanofibrous scaffolds as a model for in vitro skin tissue regeneration.* J Mater Sci Mater Med, 2012. 23(6): pp. 1511–1519.

[41] Venugopal, J. and S. Ramakrishna, *Biocompatible nanofiber matrices for the engineering of a dermal substitute for skin regeneration.* Tissue Eng, 2005. 11(5–6): pp. 847–854.

[42] Veleirinho, B., et al., *Nanofibrous poly(3-hydroxybutyrate-co-3-hydroxyvalerate)/chitosan scaffolds for skin regeneration.* Int J Biol Macromol, 2012. 51(4): pp. 343-350.

[43] Silva, S.Y., et al., *Double blind, randomized, placebo controlled clinical trial for the treatment of diabetic foot ulcers, using a nitric oxide releasing patch: PATHON.* Trials, 2007. 8: p. 26.

[44] Demircan, M., Cicek, T. and M.I. Yetis, *Preliminary results in single-step wound closure procedure of full-thickness facial burns in children by using the collagen-elastin matrix and review of pediatric facial burns.* Burns, 2015. 41(6): pp. 1268–1274.

[45] Liu, Y., et al., *Electrospun nanofibrous sheets of collagen/elastin/polycaprolactone improve cardiac repair after myocardial infarction.* Am J Transl Res, 2016. 8(4): pp. 1678–1694.

[46] Wang, S., et al., *Fabrication of small-diameter vascular scaffolds by heparin-bonded P(LLA-CL) composite nanofibers to improve graft patency.* Int J Nanomed, 2013. 8: pp. 2131–2139.

[47] Cheng, Y., et al., *Collagen functionalized bioactive nanofiber matrices for osteogenic differentiation of mesenchymal stem cells: bone tissue engineering.* J Biomed Nanotechnol, 2014. 10(2): pp. 287–298.

[48] Bou-Akl, T., Miller, R. and P. VandeVord, *Collagen nanofibers induce spontaneous osteogenic differentiation of rat bone marrow stromal cells.* Jacobs J BMSC Res, 2015. 1(1): p. 003.

[49] Wang, W., et al., *Aligned nanofibers direct human dermal fibroblasts to tenogenic phenotype in vitro and enhance tendon regeneration in vivo.* Nanomedicine (Lond), 2016. 11(9): pp. 1055–1072.

[50] Martin, J.T., et al., *A radiopaque electrospun scaffold for engineering fibrous musculoskeletal tissues: Scaffold characterization and in vivo applications.* Acta Biomater, 2015. 26: pp. 97–104.

Chapter 7

Electrospinning for Drug Delivery Systems: Drug Incorporation Techniques

Cornejo Bravo José Manuel,
Villarreal Gómez Luis Jesús and
Serrano Medina Aracely

Additional information is available at the end of the chapter

http://dx.doi.org/10.5772/65939

Abstract

Electrospinning is a very versatile techniqueused for many purposes, such as tissue engineering, textiles, air and water treatment filter, solar cells, and drug delivery systems, among others. This method is cheap, easy to handle, reproducible when ambient parameters are controlled, and can be used for many formulations. The objective of this review is to enlist and emphasize the advantages and disadvantages of different methods for incorporating therapeutic drugs in a drug delivery system with electrospinning. The importance of the research to create new and innovative drug carriers is high, because of their efficiency of transporting the bioactive agent to the target zone, avoiding secondary effects in the body. Nanofibers and nanoparticles have become an important strategy in pharmacology due to their physicochemical and biocompatible properties useful for this purpose. Among the techniques compared are blending coaxial, emulsion and surface modification electrospinning, followed by electrospray and coaxial electrospray. The present review concludes that every technique has advantages and disadvantages and, not all drugs can be loaded with any method, the strategy used will depend on the drug's physicochemical properties, target zone, polymeric characteristics, and required drug release rate. This chapter will serve as a starting point for when to choose one of the drug incorporation techniques mentioned.

Keywords: drug delivery system, drug incorporation, electrospinning

1. Introduction

The electrospinning technique has been widely used for drug delivery system approach. This versatile procedure uses an electrical field to create nanofibers from a conductive solution. These solutions can be prepared from polymeric or composite materials [1–3].

Nanofibers fabricated by the electrospinning approach have been reported to be important in the pharmaceutical industry thanks to its properties, such as degradability, high surface area, variable porosity, and manageability; hence, these structures have been studied as drug delivery systems. The polymers used in this technique usually facilitate the incorporation of different drugs thanks to their chemical composition and physicochemical characteristics. For example, poly(vinyl pyrrolidone) (PVP) scaffolds improve the solubility of some hydrophobic drugs because PVP nanofibers help in the dispersion of these drugs [4].

Drug delivery systems that use the electrospinning strategy possess better control and predictability on drug delivery than the conventional methods. Such is the case of itraconozale used for the treatment of tinea pedis and other infections. This drug has been successfully grafted in electrospun nanofibers showing a lineal dependence to a square root of release rate, indicating that this drug delivery system follows a Fickian release kinetics [4].

In the case of mexofin, a cephalosporin antibiotic used to treat bacterial infections or preventing bacterial infections before, during, or after certain surgeries, it has been reported to be loaded to poly(D,L-lactid acid) (PDLLA) scaffolds showing a complete release of the drug within 24 h with a bulk release within the first 3 h. The researchers proposed that the surface deposition and drug accumulation has an important effect on drug delivery performance of electrospun nanofibers [5].

Other studies reported drug delivery systems using poly(ethylene-co-vinyl acetate) (PEVA)/poly(lactic acid) (PLA) nanofibers loaded with the polyketide antibiotic tetracycline, well known to treat periodontal diseases. Comparing the PEVA scaffolds with the PLA/PEVA nanofibers, the first system proved to have a higher release, delivering 65% of the loaded drug in 120 h [6].

Additionally, not only drugs can be loaded in electrospun nanofibers or microfibers, but also biological agent can be controlled and released with this system. For this approach, it has been reported that nanofibers with lower diameters are effective for delivering genes, proteins, and enzymes in postsurgical treatment of glioma cells [4].

2. Advantages and disadvantages of nanofibers as drug carriers

Material uses for drug delivery systems need to be biocompatible, biodegradable, permit drug loading, possess mass transfer properties, and respond to stimuli, among others. Drugs can be loaded into the polymeric nanofibers from antibiotics and anticancer agents to biomolecules such as DNA, RNA, and proteins [2].

http://dx.doi.org/10.5772/65939

Among the advantages of electrospun nanofibers as drug carriers is the high surface-to-volume ratio, which can accelerate the solubility of the drug in the aqueous solution and improve the efficiency of the drug. Surface morphology and structure of the nanofibers are important factors for controlling the releasing rate and amount of the drug. Furthermore, biodegradable polymers would protect drug from corrosion of gastric acid and enzyme, maintaining the bioactivity of the drug. On the other hand, nanofibers scaffolds can be used as templates for the production of conductive drug-loaded polymer systems [7].

It can be found in the literature that electrospinning is a cost-effective technique capable of producing long and continuous nanofibers. Also, it is useful for the production of aligned nanofiber and tailorable mechanical properties. Despite this, the method possesses several limitations as production of large nanometer to micron-scale fiber requires the use of organic solvents and it is not possible to control the 3D pore structure [8].

Compared with other methodologies, electrospinning is an easy and friendly technique. One-step top-down process in the fabrication of nanoscale fibers is not limited in the coprocessing of polymeric mixtures or chemical cross-linking. The scaffolds made by electrospinning possess various properties such as charged surfaces areas, adequate porosity, elasticity, strength, and weight. The disadvantages of the other techniques for the production of nanofibers such as template synthesis, drawing, phase separation, and self-assembly are material limitation, time consuming, and complex processing [9].

One of the important things to recognize about electrospinning is the demonstration of an enhanced drug release rates using nanofiber formulation than those from original drug substance reported in *in vitro* and *in vivo* studies [7].

3. Electrospun polymeric nanofibers as drug delivery systems

A great variety of polymers have been used to form nanofibers, such as poly(L-lactic acid) (PLLA), poly(L-lactic acid)/hydroxyapatite (PLLA/HA) [10, 11], poly(glycolic acid) (PGA), poly(caprolactone) (PCL), poly(carbonate), poly(urethane) (PU), poly(ethylene glycol) (PEG) [1], poly(D,L-lactide-co-glycolide) (PLGA), PLGA/gelatin [7], poly(lactic acid) (PLA), poly(ε-caprolactone) (PCL), poly(ethylene oxide), and poly(L-lactide-co-caprolactone) [12], among others (**Table 1**). But not all of them can be used for drug delivery systems.

Many drugs have been loaded into nanofibers using different methods. Among these drugs, chromazural B was successfully loaded into 5, 10, 15, 20-tetraphenyl-21H, 23H-porphine tetrasulfonic acid tetrasodium (TPPS) fibers; in this work, the scientists concluded that the fiber diameter was inversely proportional to the drug release rate [18] (**Table 2**).

Electrospun polymeric nanofibers	Drug loaded	Type of drug	Ref.
Poly(vinyl alcohol) (PVA)	Sodium salicylate, diclofenac sodium, naproxen (NAP), and indomethacin (IND)	Freely soluble in water, sparingly soluble in water, both insoluble in water, respectively	[13]
Poly(ethylene-co-vinyl acetate) (PEVA)	Actisite®	Antibiotic originally isolated from *Streptomyces aureofaciens*	[14]
Poly(vinyl alcohol)/poly(vinyl acetate)	Ciprofloxacin HCl (CipHCl)	Antibiotic belong to a group of drugs called fluoroquinolones	[15]
Poly(D,L-lactide-co-glycolide) (PLGA), PLGA/gelatin	Fenbufen	Nonsteroidal anti-inflammatory drug	[7]
Poly(ε-caprolactone) (PCL)	Naproxen (NAP)	Non steroidal anti inflammatory drug (NSAID) of the propionic acid class that relieves pain, fever, swelling, and stiffness	[16]
Poly(vinyl alcohol) (PVA) and sodium alginate (Na-Alg)	Insulin	Pancreas´ hormone that allows the body to use sugar	[17]

Table 1. Nanofibers as drug delivery systems.

Fibers/particles	Drug loaded	Observations	Ref.
TPPS nanofibers[a]	Chromazural B	Smaller fibers exhibited rapid drug release in initial stage No significant difference between the total amount of released drug and the fiber diameter or type of drug, fiber diameter between 150 and 290 nm	[18]
PNIPAAm nanofibers[b]	Ketoprofen (KET)	Exhibited fibers largely smooth and cylindrical, with no phase separation	[3]
Gliadin nanoparticles	Cyclophosphamide anticancer drug	Particles are homogeneous and have a narrow distribution, nanoparticles diameter between 218.66 ± 5.1 nm	[19]
PCL[c] and PHBV[d] nanofibers	Metformin hydrochloride (MH) or metoprolol tartrate (MPT)	Emulsion electrospun nanofibers significantly alleviated the burst release and produced a sustained release of drugs compared to the blended electrospun nanofibers	[3]

[a]TPPS: 5,10,15,20-tetraphenyl-21H, 23H-porphine tetrasulfonic acid tetrasodium.
[b]PNIPAAm: poly(N-isopropyl acrylamide).
[c]PCL: poly(ε-caprolactone).
[d]PHBV: poly(3-hydroxybutyric acid-co-3-hydroxyvaleric acid).

Table 2. Characteristics of fibers and particles for drug delivery applications.

4. Drug incorporation techniques using electrospinning

Nanofibers are not the only product of electrospinning device useful for drug delivery systems. It can also be applied to the electrohydrodynamic (EHD) technique, which allows the production of nanoparticles of different shapes and sizes. This method employs several techniques such as blending, coaxial process, and surface modulation, which permits the

http://dx.doi.org/10.5772/65939

incorporation of any drug, DNA, and growth factors, depending on the treatment needed to reach the proper material characteristics (**Figure 1**). Depending of the bioactive molecule desirable to be loaded, the polymeric or composite system is carefully chosen, with the purpose of the preservation of the therapeutic effect [12].

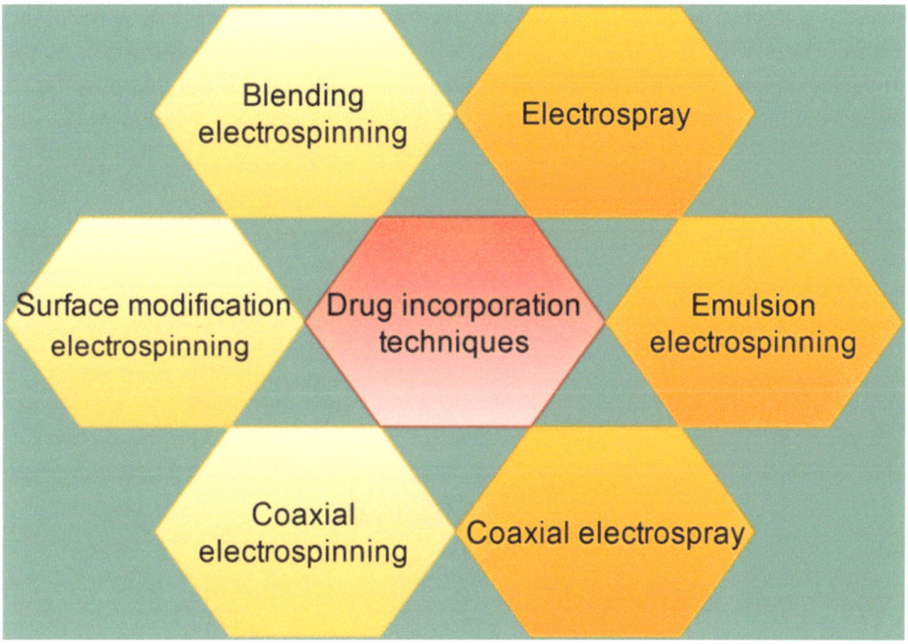

Figure 1. Drug incorporation techniques by electrospinning device. Adapted with permission from Ref. [11].

4.1. Blending electrospinning

The use of polymeric blend improves the equilibrium between mechanical and physicochemical properties of the drug-loaded nanofibers. Also, it effectively increases the formulation design for drug release, where the release rate can be manipulated by altering the proportion of polymer in the blended solution [20].

When using the blending electrospinning method, drug encapsulation is achieved through electrospinning in a single step, because drugs are dissolved or dispersed in the polymeric solution (**Figure 2**). The interaction between polymer solution and drugs is affected by the physicochemical properties of the polymers, because these characteristics act as factors that determine the efficiency in drug encapsulation, drug dispersion into the fibers, and the release rate. The isolated release of the drug into the solution can be triggered by the insufficient solubility of the drug in the polymeric solution, where the drug molecules can migrate to a nearby fiber's surface during the electrospinning process. Researchers have emphasized the importance of the equilibrium between hydrophilic and hydrophobic properties among drugs and polymers when blending electrospinning is used. For example, thanks to the hydrophobic nature of some polymers, the lipophilicity of drugs becomes easier to get a homogeneous solution, and vice versa. For example, polyester polymers, which are hydrophobic, interact very well with the hydrophobic drug rifampicin and paclitaxel, and gelatin, PEG, and PVA, which are hydrophilic polymers, can dissolve hydrophilic drugs such as doxorubicin [21].

With blending electrospinning fibers are obtained with a single phase only. If the objective is to produce fibers with core-shell structure, which protect the labile biological agents and growth factors, blending electrospinning is not the proper method. For this purpose, coaxial and emulsion Electrospinning can be used [21].

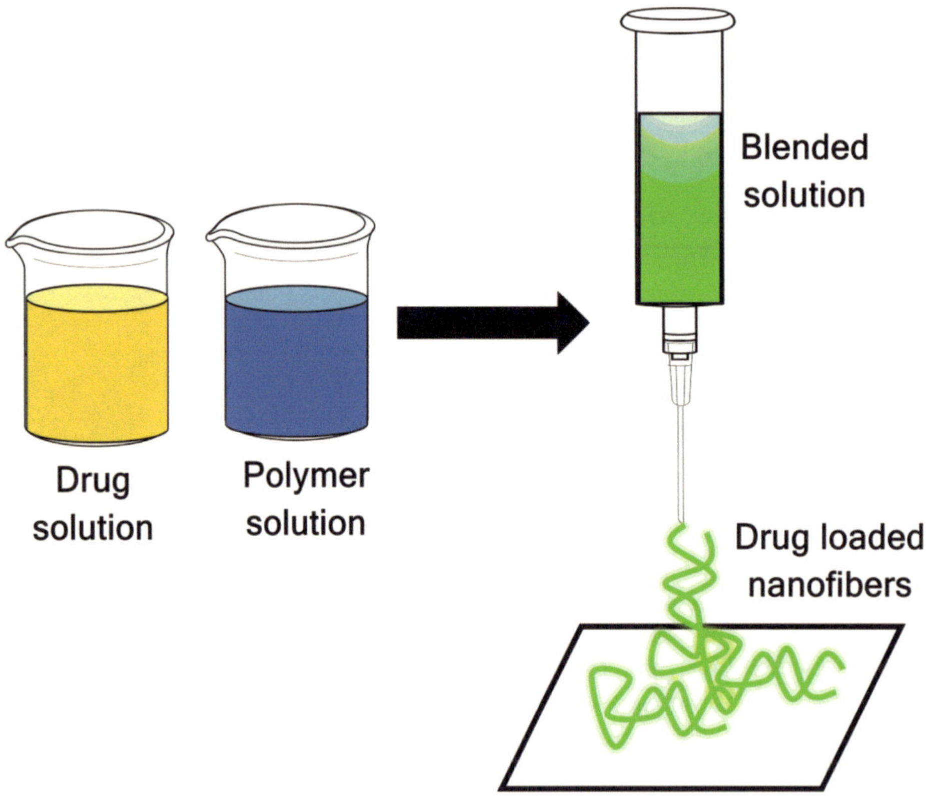

Figure 2. Blending electrospinning principle.

Several reports can be found in the literature where blending electrospinning is used for the loading of drug in a polymeric delivery system. A research, conducted by Lu et al., produced fibers of poly (N-isopropyl acrylamide) (PNIPAAm), which is well known as a smart polymer because it responds to pH and temperature stimuli, mixed with ethyl cellulose (EC), and it was successfully blended by electrospinning, and ketoprofen (KET), a nonsteroidal anti-inflammatory drug, was added into the system. The resulting complexes exhibited fibers that are largely smooth and cylindrical, with no phase separation. In this investigation, it was found that the drug was present in its amorphous physical form in the fiber matrix, and significant intermolecular interaction between KET and the polymeric scaffold was observed. These systems were not toxic and biocompatible in cell culture [3].

4.2. Coaxial electrospinning

The main purpose of coaxial electrospinning is to obtain fibers with core-shell structure. This technique can be used to obtained fibers with specific drugs encapsulated in the core of the fibers, which lead to a sustained and controlled drug release (**Figure 3**). These kinds of fibers present a high surface area and three-dimensional network. Proteins, growth factor, antibiotics, and other biological agents have been successfully loaded into the coaxial fibers for drug delivery purposes. One of the main advantages of this technique is that the core-shell structure gives protection to the loaded compound and the bioactivities of drugs, for example, remain intact [9].

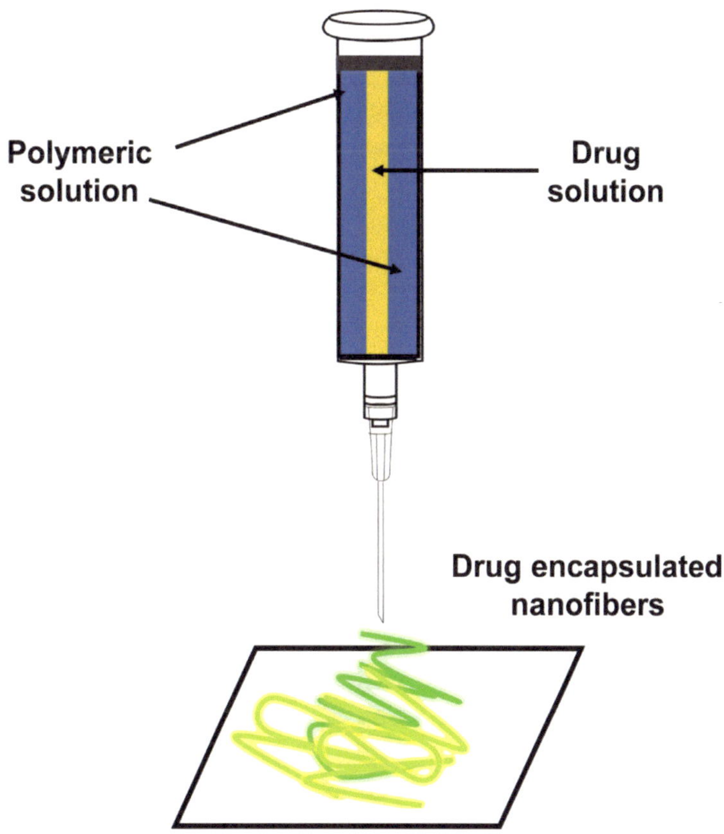

Figure 3. Coaxial electrospinning principle.

Among the coaxial electrospinning method benefits is the enhancing of biomolecule functionality, by having it into the inner jet, while the electrospinning process is working and the polymeric solution is in the outer jet giving protection to the biomolecule. In this technique, the polymeric shell helps to avoid the direct contact of the biomolecule with the external environment. The core-shell system improves the sustained release of drugs and also allows the bioability of unstable biological agents to be maintained [21].

4.3. Emulsion electrospinning

Emulsion electrospinning is a flexible and potential technique for the encapsulation of several drugs into nanofibers [3] and is one of the most important methods for preparing core-shell electrospun nanofibers in a cost-effective and efficient manner (**Figure 4**) [9].

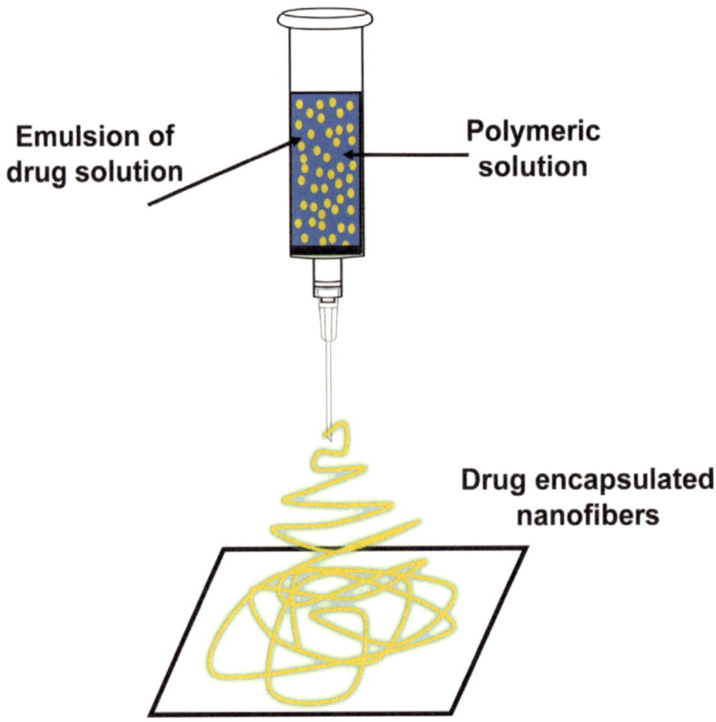

Figure 4. Emulsion electrospinning principle.

Using emulsion electrospinning fibers composed of poly (ε-caprolactone) (PCL) and poly (3-hydroxybutyric acid-co-3-hydroxyvaleric acid) (PHBV) were loaded with metformin hydrochloride (MH) or metoprolol tartrate (MPT). In this study, emulsion electrospinning demonstrated to be a better technique than blending electrospinning, especially in the modulation of the drug release rate by regulating the oil phase and water phase of the emulsions for obtaining the desired drug release. Between the two polymers tested, PCL showed better drug delivery properties than PHBV [3].

In the emulsion electrospinning method, the oil phase is created by the emulsion of the drug or aqueous protein solution in the polymer solution, followed by electrospinning. When the

http://dx.doi.org/10.5772/65939

drug to be load has a sufficient low molecular weight, the biomolecule-loaded phase can be distributed within the fiber or a core-shell fibrous structure could be configured as macromolecules amalgamate in the aqueous phase. The advantage of emulsion electrospinning against blending electrospinning is the elimination of the need for a common solvent as the drug and the polymer are dissolved in applicable solvents. Subsequently, a number of formulations of hydrophilic drugs and hydrophobic polymeric solution can be used while maintaining minimal drug contact with the organic solvent during the procedure [21].

4.4. Surface modification electrospinning

In this strategy, thanks to the electrospinning approach, a particular conductive surface can be chemically altered and changed, with the purpose of modifying the external properties of the coated device, by incorporating certain molecules that can camouflage the surface by offering a similar environment than the tissue that will surround the implanted material. Usually this strategy is applied to avoid fast initial burst release and slow the rate of immobilization of the biological molecules on a particular surface. In addition, with a good electrospinning system and with a well-standardized method to create electric field inside a camera, it is possible to coat 3D surfaces with nanoparticles or homogeneous surfaces (**Figure 5**) [21].

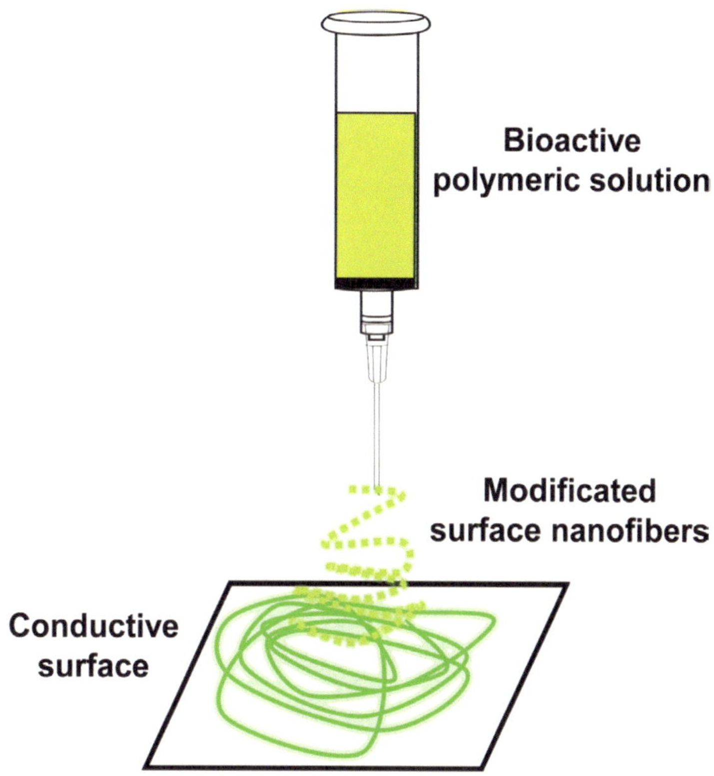

Figure 5. Surface modification electrospinning principle.

4.5. Electrospray

Electrospray is one of the most effective methods for the synthesis of nanoparticles and nanospheres. This technique is the simplest of all the drug incorporation methods. In this process, the liquid emerging from the nozzle into the electrical field forms the Taylor cone because of the surface tension. Once the electric field increases, the Taylor cone breaks into highly charged droplets, creating suitable conditions for the formation of nanoparticles or microparticles. Solid particles are formed by solvent evaporation (**Figure 6**). Some of the parameters that need to be taken into account are the needle gauge diameter, flow rate, voltage, and distance from the needle to the conductive collector, promoting the right incorporation of the drug [19].

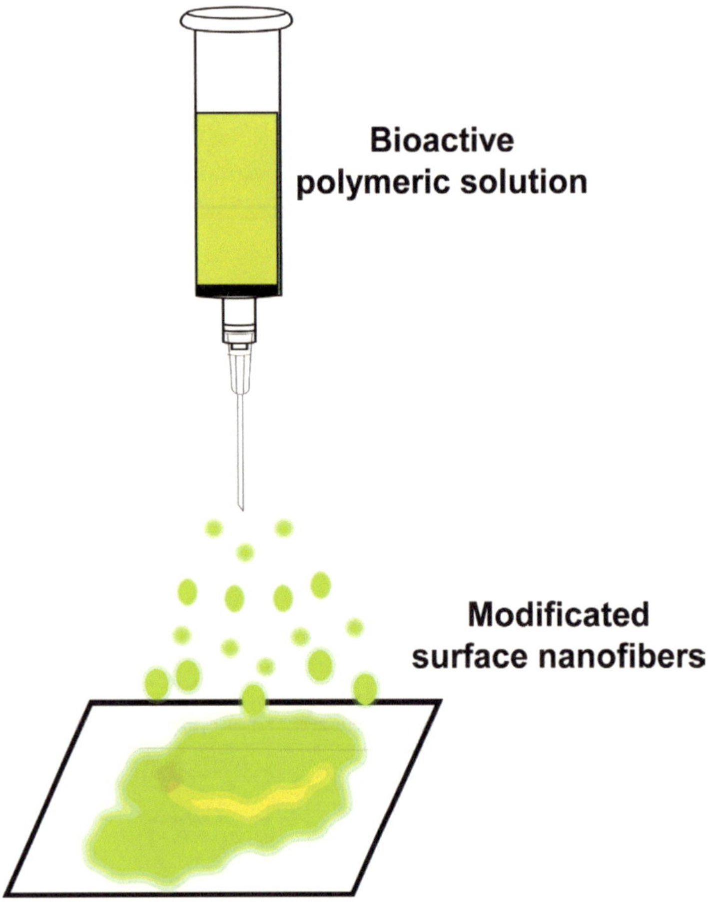

Figure 6. Electrospray principle.

Electrosprayed nanoparticles can be useful for biological, medicinal, and pharmaceutical applications; hence, it is zero dimensional in nature. The advantages of electrospraying include increased scalable synthesis, reproducibility, and high encapsulation efficiency [19].

In addition, with this technique nanoparticles can be loaded with pharmaceutical molecules like drugs and biomolecules acting as particular transporters due to its active surface absorption, binding capacity, and complexation with drugs and bioactive molecules. Moreover, the particles of a nanoscale size are important in therapeutic treatments, because they improve the drug carrier rate, specificity, adhesion, and reactivity [19].

For this strategy, the use of nanoparticles of natural gliadin loaded with cyclophosphamide anticancer drug has been reported for the treatment of retinoblastoma and certain cancers. In this study, over 72% of drug loading was accomplished; researchers reported homogeneous nanoparticles with average diameters of 218 ± 5.1 nm [19].

4.6. Coaxial electrospray

Coaxial electrospray allows the production of multilayer particles with sizes ranging between 10 and 100 μm by using a high electric field between coaxial capillary needle and ground. In this technique, the resultant electrical shear stress elongates the core and the shell liquid menisci at the needle outlet to form the "Taylor cone"; after this phenomenon, the jet of the liquid elongates enough until it is broken into multilayer droplets owing to the electrohydrodynamic forces (**Figure 7**) [22].

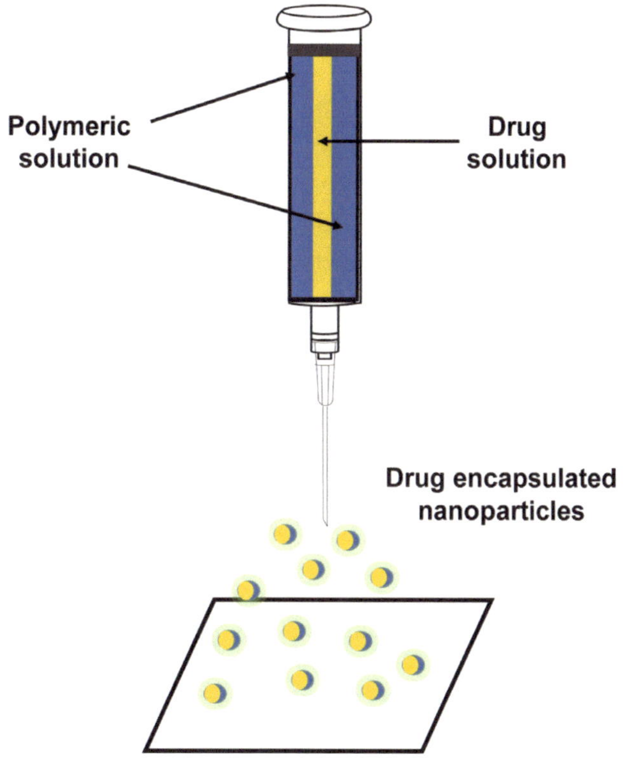

Figure 7. Coaxial electrospray principle.

This method achieves high encapsulation rate, precise control of the core-shell structure, and protection of the fragile therapeutic cargos from process-induced denaturation and aggregation. Also, this process is scalable for mass production of drug-loaded microparticles and nanoparticles [22].

Moreover, coaxial electrospray leads to a microencapsulation and a nanoencapsulation of drugs into polymeric particles. To perform this method it is necessary to use a coaxial capillary needle to deliver two liquids independently. Several liquid materials can be used including glycerol, distilled or deionized water, ethanol, ethylene glycol, etc. For the synthesis of multilayer macroparticles and nanoparticles with a hard shell, polymeric solution, such as PCL, PLA, PLGA, PMMA, PE, among others, can be used [22].

In general, all methods for drug incorporation using electrospinning possess its advantages and limitations (**Table 3**).

Method	Advantages	Disadvantages	Ref.
Blending electrospinning	Improves the tunability of the physicochemical and mechanical properties of the drug-loaded fibers. Benefits the development of controlled drug release formulations, for which the release rate can be modified by altering the ratio of the polymers in the blend	A clear understanding of the phase behavior of the processed polymer blend is essential	[20, 21]
Coaxial electrospinning	Enhanced biomolecule functionality. The core ingredient is shielded by the shell polymer avoiding direct contact to the biological environment	Need a special syringe tip	[21]
Emulsion electrospinning	No need of a common solvent as the drug and the polymer are dissolved in applicable solvents	Not all drugs can be loaded by this method	[21]
Surface modification electrospinning	Resolve the issues of large initial burst release and short release time as the biomolecules are surface immobilized	Depend of the nature of polymers and drugs	[21]
Electrospray	Easy to control the operation parameters. Fast preparation and one-step technique. This involves simple ideology. This technique can be able to extend for bulk production	This technique may induce some macromolecule degradation due to the stress involved in the operation parameters (ex: thermal stress in drying, shear stress in the nozzle)	[19]
Coaxial electrospray	High encapsulation efficiency, effective protection of bioactivity and uniform size distribution	Process control in coaxial electrospray is challenged by the multiphysical nature of the process and the complex interplay of multiple design, process and material parameters	[22]

Table 3. Advantages and disadvantages of drug incorporation methods using electrospinning.

5. Conclusion

This chapter summarized reported drug incorporation techniques developed from electrospun nanofibers and nanoparticles, from polymers used to a comparison between the techniques. Coaxial electrospray and coaxial electrospinning can be used when a core-shell structure is need, for example, in cases when the therapeutic agent is sensible to the environment. Blending and emulsion electrospinning do not need special equipment and are the simplest methods to incorporate drugs into nanofibers. Finally, surface modification is necessary when the rampage effect needs to be avoided and more perdurable release rate is desired. All techniques are useful, versatile, cheap, and easy for the incorporation of drug in a drug delivery system. The method must be chosen depending on the nature of the therapeutic agents.

Author details

Cornejo Bravo José Manuel[1,2*], Villarreal Gómez Luis Jesús[1,2,3] and Serrano Medina Aracely[1,4]

*Address all correspondence to: jmcornejo@uabc.edu.mx

1 Autonomous University of Baja California, Tijuana, Baja California, Mexico

2 Faculty of Chemical Sciences and Engineering, Autonomous University of Baja California, Tijuana, Baja California, Mexico

3 School of Sciences of the Engineering and Technology, Autonomous University of Baja California, Tijuana, Baja California, Mexico

4 Faculty of Medicine and Psychology, Autonomous University of Baja California, Tijuana, Baja California, Mexico

References

[1] Argawal S, Wendorff JH, Greiner A. Use of electrospinning technique for biomedical applications. Polymer. 2008;49(26):5603–5621. DOI: 10.1016/j.polymer.2008.09.014.

[2] Sill TJ, von Recum HA. Electrospinning: applications in drug delivery and tissue engineering. Biomaterials. 2008;29(13):1989–2006. DOI: 10.1016/j.biomaterials.2008.01.011

[3] Hu J, Prabhakaran MP, Tian L, Ding X, Ramakrishna S. Drug-loaded emulsion electrospun nanofibers: characterization, drug release and in vitro biocompatibility. RSC Advances. 2015;5(21):100256–100267. DOI: 10.1039/C5RA18535A

[4] Gupta KC, Haider A, Choi Y, Kang I. Nanofibrous scaffolds in biomedical applications. Biomaterials Research. 2014;18(5):1–11. DOI: 10.1186/2055-7124-18-5.

[5] Nikkola L, Morton T, Balmayor ER, Jukola H, Harlin A, Redl H, van Griensven M, Ashammakhi N. Fabrication of electrospun poly(D,L-lactide-co-glycolide)80/20 scaffolds loaded with diclofenac sodium for tissue engineering. European Journal of Medical Research. 2015;20(1):54. DOI: 10.1186/s40001-015-0145-1.

[6] Kenawy ER, Bowlin GL, Mansfield K, Layman J, Simpson DG, Sanders EH, Wnek GE. Release of tetracycline hydrochloride from electrospun poly(ethylene-co-vinylacetate), poly(lactic acid), and a blend. Journal of Controlled Release. 2002;81(1–2):57–64. DOI: http://dx.doi.org/10.1016/S0168-3659(02)00041-X.

[7] Meng ZX, Xu XX, Zheng W, Zhou HM, Li L, Zheng YF, Lou X. Preparation and characterization of electrospun PLGA7 gelatin nanofibers as a potential drug delivery system. Colloids and Surfaces B: Biointerfaces. 2011;84(1):97–102. DOI: 10.1016/j.colsurfb.2010.12.022

[8] Rathinamoorthy R. Nanofiber for drug delivery system–principle and application. P. T. J. 2012, 61, 45–48.

[9] Lu Y, Huang J, Yu G, Cardenas R, Wei S, Wujcik EK, Guo Z. Coaxial electrospun fibers: applications in drug delivery and tissue engineering. Wiley Interdisciplinary Reviews: Nanomedicine and Nanobiotechnology. 2016;8(5):654–677. DOI: 10.1002/wnan.1391

[10] Villarreal-Gomez LJ, Cornejo-Bravo JM, Vera-Graziano R, Grande D. Electrospinning as a powerful technique for biomedical applications: a critically selected survey. Journal of Biomaterials Science-Polymer Edition. 2016;27(2):157–176.

[11] Villarreal-Gomez LJ, Vera-Graziano R, Vega-Rios MR, Pineda-Camacho JL, Almanza-Reyes H, Mier-Maldonado PA, Cornejo-Bravo JM. Biocompatibility evaluation of electrospun scaffolds of poly(L-lactide) with pure and grafted hydroxyapatite. Journal of the Mexican Chemical Society. 2014;58(4):435–443.

[12] Zamani M, Prabhakaran MP, Ramakrishna S. Advances in drug delivery via electrospun and electrosprayed nanomaterials. International Journal of Nanomedicine. 2013;8(1):2997–3017.

[13] Pattama T, Uracha R, Pitt S. Drug-loaded electrospun mats of poly(vinyl alcohol) fibres and their release characteristics of four model drugs. Nanotechnology. 2006;17(9):2317.

[14] Yu D, Zhu L, White K, Branford-White C. Electrospun nanofiber-based drug delivery systems. Health. 2009;1(1):67–75. DOI: 10.4236/health.2009.12012.

[15] Jannesari M, Varshosaz J, Morshed M, Zamani M. Composite poly(vinyl alcohol)/poly(vinyl acetate) electrospun nanofibrous mats as a novel wound dressing matrix for controlled release of drugs. International Journal of Nanomedicine. 2011;6(1):993–1003. DOI: http://doi.org/10.2147/IJN.S17595